The time when these 'new' inventions were being described was in the reign of Queen Victoria of England. Here the nostalgia of that innocent and interesting period is brought out in authentic detail. Actual engravings from the scientific and popular press of the period illustrate the contemporary accounts of the 'latest' mechanical and electrical wonders.

'The Telephone', for example (an apparatus for the electrical transmission of distinctly articulate sounds to great distance), is here shown in elegantly engraved illustrations, with authoritative notes on the use of the instrument. There is also news of the latest trends in Photography (a portable 5 lb. Hand Camera, and Sensitive Material on Rollers) and there is even talk of the remarkable shape of things to come, like the Submarine Railway Between France and England, and the new experiments in Aerial Locomotion. There are also details of the Patent Impulsoria and the Aërophon, not to mention the Pedal Zephyrion and a project for a Machine for Sensational Emotions.

No aspect of daily life is neglected; also included are the Combined Chest Expander and Skirt Supporter, and Mr Raimes' Apparatus for the Toilet. The review is completed with an outline of a novel theory: Aerial Propulsion by Explosion or Impulse—an entirely new concept, utilizing dynamite as fuel, for crossing the Atlantic in as little as a Few Hours.

NEW INVENTIONS.

NEW INVENTIONS.

A COMPREHENSIVE SURVEY OF

Scientific and Technical Progress

in the

ARTS, SCIENCES AND MANUFACTURES

as published during
THE REIGN OF HER MAJESTY.

EDITED AND PRESENTED
by

MAURICE RICKARDS.

New York

HASTINGS HOUSE, PUBLISHERS.

FIRST AMERICAN EDITION PUBLISHED IN 1969 BY
HASTINGS HOUSE PUBLISHERS, INC.,
10 EAST 40TH STREET, NEW YORK 10016.
THIS EDITION © 1968, MAURICE RICKARDS.
S.B.N. 8038-5012-3.
LIBRARY OF CONGRESS CATALOG CARD No. 69-16544
PRINTED AND BOUND IN THE REPUBLIC OF IRELAND

CONTENTS.

THE PNEUMATIC LETTER AND PARCEL CONVEYANCE: THE DISC IN THE ENGINE HOUSE (SEE PAGE 33.)

CHAPTER I.

INTRODUCTION: NO IMPOSSIBLE INVENTIONS.

WHEN someone suggests a bridge over Behring Straits, or a tunnel under the Straits of Dover, the idea is laughed at, and the swift and easy word "impossible" finds utterance. So, also, when a crank says we can build a vessel that will make the voyage from America to England in a day, or perhaps a trifle more, he is laughed at and told that it is all impossible. And the other crank, who is getting up balloons that will sail in any direction, and can be driven by some high power that will force great speed, he, too, is right merrily laughed at as engaged on the "impossible." We merely cite these for examples, for we have now in our mind that other fellow, who is going to give a universal and cheap supply of illuminating gas and fuel by merely decomposing water. To be sure, this latter party has come along too often, and done too little to ensure, at once, his status as a real inventor having a practical head on him. All the same, we take stock in all these inventions and others of the same extravagant sort. We hardly think any decent human want will be left unsupplied, even if at first blush the vagary does seem to run wild and the grasp is far reaching. Man's skill and ingenuity appear in the light of past experience almost equal to an impossibility. The real marvel too, is that these impossible things, when once done, fall so easily into the list of common sort of things, that, after all, did not require so much real wit to conceive and perfect. The telegraph wire under the Atlantic was once thought a big thing, but the finding and taking up in mid ocean of the ends of one of the earlier broken cables, and then and there uniting them was and remains a marvel, even if the deed is not often recounted.

Edison's phonograph and its possibilities, of which as yet we have only very dim understanding, is close on to the marvellous. So, also, the wonders of photography, which we have yet hardly begun to understand, and the equally great and wonderful impossibilities of electricity, of which we are about wholly in ignorance. In fact, it seems only a question of time, when the majority will be of the opinion that there is no need of this word "impossible" in any language. Certainly none for a word as applied to inventions.

Every one is a potential inventor, especially if he have an acquaintance with science or manufacture. Whatever a man's occupation, he must daily find himself called upon to do or to suffer many things from which he would gladly emancipate himself. The proverb says, "There is a remedy for every ill but death," and seeing how many ills there are, the opportunities for devising remedies are not only numberless but they are present to all. The unenterprising bear with patience the inconveniences that surround them, but those of active mind busy themselves in devising expedients to lighten the burden of life, and look for their reward under the provisions of the patent law. The inventor is the greatest benefactor of the human race, and especially of that part of it which is indigent; he is the real friend of the poor man, and, indeed, almost his only friend.

In this review we present a conspectus of enterprise, a record of the inventor's contribution to the comfort of all classes, rich and poor alike. The record serves, we venture to opine, not only as history, but as inspiration for invention still to come.

CHAPTER II.

Part I: Of Interest to Ladies.

The Aërephon.

The Aërephon is introduced to the musical public by Mr. Arthur S. Denny, who brought before the public, at the Crystal Palace, an instrument of similar principle but much less perfect in its arrangements. The instrument, under the title of "Calliope" was brought to the exhibition in an imperfect condition. It was Mr. Denny's object to secure the interest of the musical and scientific gentlemen of Great Britain, in order that he might be enabled to bring the invention to a higher state of perfection. We learn that he is indebted to Messrs Horne and Thornthwaite, of Newgate-street; Mr. Henry Willis, the organ-manufacturer; Mr. Henry Distin, the musical instrument manufacturer, and others, for many valuable suggestions, which have enabled him to mature and apply important improvements. Mr. S. B. Simpson, the enterprising proprietor of Cremorne Gardens, ever on the alert to gratify a laudable curiosity on the part of the public to see anything new in science or music, has arranged with Mr. Denny to have this instrument brought before the public at his popular place of amusement, where those interested in such matters may always have an opportunity of witnessing its performances. Every night, we understand, the powerful notes of the Aërephon will furnish music for one or more dances. Dancing by steam is novel indeed; but, judging from the private performances of the instrument we see no reason why it should not furnish good music for dancing.

This truly wonderful invention far exceeds any expectation of its power and capacity. Although worked by steam it is capable of producing the highest swell or the lowest symphony; and, while its loud, sonorous tones may from its present position be heard from Hungerford-bridge, they can be so modulated and governed as to be made agreeably sweet, and but moderately audible at a distance of one hundred yards.

The mechanism of the instrument is simple in its nature and construction. The steam, which operates instead of air on the brass pipes, is confined in a chamber on which the pipes are arranged. Connected with each pipe is a valve of peculiar construction and very easy action. The valves are opened by means of pianoforte keys, attached by wires and closed by springs at their backs, assisted by a slight force of steam. One of the keys, being pressed upon, causes the corresponding valve to open, thereby admitting the steam to the pipe with which it is connected, and producing its appropriate sound. There are twelve of the large bass-trumpets on one side, and twenty-two soprano pipes on the other.

An Improved Waterproof.

Mr. Robert Thomson, of Aberdeen, is the patentee of an improved armhole for waterproofs. This improvement in waterproof capes has not come a bit too soon. The old style of cape with the wide flap—buttoned top and bottom, and often in the middle—never found favour, owing to its extreme liability to tear. Many ideas have been put forward as a substitute, such as Dolman

MR. ARTHUR DENNY'S REMARKABLE NEW STEAM INSTRUMENT—THE AËREPHON.

and Princess shapes with sleeves, but these have a limited demand and must always have so, as such goods require fitting and can only be worn upon ladies in their figure. It has long been decided that the circular cape stands alone for simplicity and convenience, and this contrivance for rendering the cape serviceable and untearable must be hailed with some considerable satisfaction. We are informed that since its introduction many hundreds have been sold, and we venture to say that time will establish it as *the* waterproof of the day. The cape has a double advantage by the arrangement of the armholes, as it also affords a rest for the hands. It is made in such a way that the whole pressure is borne by the shoulders, there being no weak points about the cape. Apart from the utility of it, its appearance is much in its favour. Our engraving (*left, above*) will help to explain the utility of the invention.

So long as it is the fashion for ladies to wear garments of the pronounced amplitude now favoured by so many of the fair sex, we do not see why the fact may not be taken advantage of to introduce an invention calculated to make it convenient for them frequently to rest from the fatigue of long standing or walking. Such, at least, we presume to be the idea of the American inventor of the device shown in the accompanying illustration (which is taken from the *Scientific American*), for which a patent has recently been issued. The transformation the style has effected in the appearance of a lady properly fitted out in walking costume is something really wonderful, and we are not surprised, therefore, that several other inventors have rushed into the same field with devices which would not otherwise have been thought of.

Speaking Tubes with Bells.

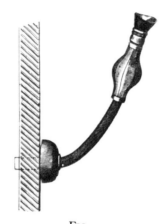

FIG. 1.   FIG. 2.

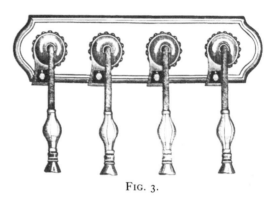

FIG. 3.

Who can tell how much time, trouble, annoyance, and even money have been saved by speaking tubes? What miles of stairs had to be traversed by servants under the old system, when bells alone existed by which those who were wanted could be summoned. During the time thus lost important opportunities have often been missed, and very substantial loss was the consequence. Mistresses and servants alike may well be grateful to the inventor of the speaking tube. It is a much humbler affair than the telegraph or the telephone, but its usefulness is undeniable. It is now in fact an indispensable adjunct of nearly all our business houses and of many homes, and until something better is devised to take its place it cannot be too highly valued. It was reserved to Mr. John W. Black, a plumber and gasfitter in Glasgow, to devise an important improvement in the speaking tube. This invention consists in the substitution of bells for the objectionable whistles now used for the purpose of attracting attention, so that there is no necessity for putting the mouth to the tube at all.

One has simply to press a little bulb at the mouthpiece and a bell is rung which signals for speaking at either end of the tube. On the signal being answered and the mouthpiece lifted to the mouth, a little ball drops away inside the seat or neck of the mouthpiece, and the tube is clear for speaking. When done with, the mouthpiece is allowed to drop from the hand, and it is again in position for signalling. We may add that the speaking is heard much more distinctly at the other end of the tube if the mouthpiece be held a little distance from the mouth. Besides being an improvement upon the old style, these speaking tube appliances are neat and even elegant in appearance when fitted up, a great contrast to the unsightly projections that one is accustomed to see. Our engravings show: fig. 1, the speaking tube in position for signalling; fig. 2, for speaking. When for house use, the bells in fig. 3 show pendulum indicators, which are set in motion whenever the bell is acted upon, and are fitted to the bells when two or more are fitted up together in the kitchen or office. The pendulum is covered with a neat case to prevent its being touched by the hand. For use on ships, where the motion of the vessel would render the pendulum indicator invalid, the indicator consists of a ticket being blown out from the side of the bell the moment it is acted upon, each ticket being numbered or lettered, as required, to show which apartment the bell is connected with. The mechanism is simple, not liable to get out of order, and withal cheap.

This improvement in speaking tubes is not the only benefit in the way of useful invention which Mr. Black has conferred upon the community. He is known as the patentee of a new w.c. pan which in its own way is likely to prove as valuable.

A Window in the Umbrella.

The difficulty of seeing one's way when carrying an umbrella at an angle is of old standing, and collisions have, ere this, happened to persons so travelling through rain or sleet. Mr. C. Howell, of 98, Shakespere-road, Herne-hill, claims to have provided for this difficulty by devising a window in the umbrella. It consists of a small oval piece of glass, with a brass frame, which is easily mounted at an appropriate level on a panel of the umbrella, while it is fixed to the silk of the umbrella by sewing through the perforated holes in the little frame. These windows can be placed in old or new umbrellas in a few minutes, and at a trifling cost, nor will they spoil the general appearance of the umbrella.

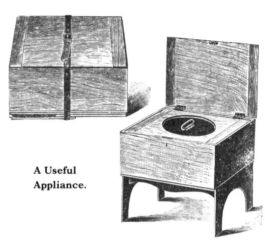

A Useful Appliance.

Those who in the pursuit of health, business, or pleasure have to make long journeys, especially abroad, know too well that what decent folk regard as an absolute essential is in many places regarded as a wild extravagance. It would seem that some peoples have made to themselves sumptuary laws which carry the social instincts of the race to an absurd extreme. To those who wish to retain the habits of our own manner of life we can heartily commend the invention figured and described here. It is in appearance when packed up like a moderately-sized bonnet box, and is therefore not objectionable as a part of ordinary travelling impedimenta. The legs fold under and are erected and secured by brackets which also fold under. When set up the box is raised to the height of 17 inches, that of an ordinary chair. A movable seat fits into the top of the box, beneath which is a japanned zinc can furnished with a handle which falls down by the side and a lid with a folding handle. The wooden lid of the box is fastened by a lock and can be taken off at pleasure. As all the internal fittings are removable the box can be used to pack articles.

A Combined Chest Expander and Skirt Supporter.

The accompanying illustrations represent a new "self-adjusting chest expanding brace and skirt support" for ladies, according to the registered design of Mrs. Pitchers, of Church Street, Godalming. Amongst the fair sex this hygienic and useful appliance should be appreciated. It can be worn with evening dress without inconvenienece or detection. The method of wearing it, as well as the advantages it embodies, will be readily seen from a glance at the illustrations.

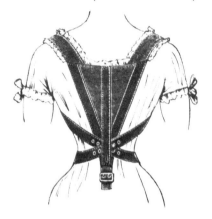 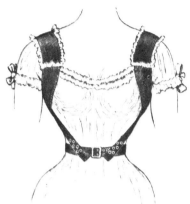

New and Instantaneous Method of Lacing Boots.

We have pleasure in drawing attention, in the accompanying illustration, to a new method of lacing boots, devised by an Irish lady, Miss Mary Sandes Hungerford, The Island, Clonakilty, county Cork. As will be seen by the sketches, the usual trouble of lacing through eyelets each time a boot is put on is avoided; the eyelet in this case is merely used as a rivet for the projecting loops, through which the lace passes up the front, and, once put in, remains in the loops.

It will readily be seen how easily boots made on this principle can be laced and unlaced. It is only necessary to pull the lace which runs up through the projecting alternate rings and the

 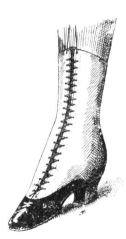

boot is securely fastened. The reverse operation is performed by drawing out the lace here and there in loops between the rings, and so loosening it sufficiently to take the boot off. This boot is made exactly like any ordinary laced boot, and the same tools and laces can be used. The strain being only on a very small portion of the lace, the latter can be changed at will, when the boot is closed (before laces are tied) by drawing it out at the vamp and adjusting.

The Pedal Zephyrion.

This ingenious invention is an apparatus whereby a fan can be actuated by the foot, leaving the hands at liberty for working or reading. Inside a hollow tube is a rod which at the lower end is connected by a hook and eye arrangement, to the end of a short lever pivoted on its centre, and which lever at the other end forms an actuating pedal.

In use the fan is fixed to a similar pivoted lever, and the motion of the foot is communicated by the fixed rod to the fan alternately raising and depressing it. Springs are placed so as to prevent jarring. We are afraid this prosaic method will not find favour with ladies when they wish to make play by the skilful use of a fan, but in the absence of an audience to captivate, we should think this plan offers an easy way of securing the comfort of a fan without employing their labour solely for that end.

An Improved Steam Washing Machine.

PEARSON & CO'S MACHINE.

The ingenuity of the inventor has been by no means confined to fields of purely masculine interest. Even in the kitchen and scullery he has made his presence felt. In the use of steam for cleaning he has helped to lighten the domestic burden. But there is still room for improvement, as the present example shows.

The Pearson Improved Steam Washing Machine is still a steam washer in its original sense, and retains all the powerful elements of steam; but what steam by itself failed to do, a small quantity of water here successfully effects, by carrying away all the dirt after dislodgment. To make this fact plain to the reader, we may state that the ordinary steam washer consists mainly of a revolving tinned copper drum, enclosed completely in a copper tank. At the bottom of the copper tank a small quantity of water is placed, which is boiled by means of an ordinary gas burner placed underneath. Round the periphery of the revolving drum, and at regular intervals are narrow slits, which, when the water in the tank boils, form a ready admittance of the steam into the interior of the drum. The clothes are placed inside the drum, and it will at once be seen, that when the drum is revolved by means of the handle attached to its axis, the clothes become thoroughly and completely impregnated by the powerful action of the steam. Messrs. Pearson & Co. have now lowered the position of this drum, so that the water itself washes into the slits which in an earlier concept were only for the admission of the steam. Inside the drum, as shown in our engraving, alongside the slits, there are attached a series of small buckets. When these buckets, revolving round with the drum, reach the water, they scoop up the boiling liquid, and carrying along until it reaches the top or highest point of the circle, they empty themselves on the top of the clothes. Thus the interior of the drum is filled with powerful steam, and the clothes have, in addition to the action of the steam, the pouring out of the boiling water upon them at the highest point of the drum.

It will be seen that this Steam Washer entirely abolishes all rubbing, brushing, peggying, and boiling of the clothes, and does not use one-fourth of the soap usually required. It is reported that it will wash a fortnight's washing for a family of eight persons in two hours, at a cost of threepence for gas and soap; can be easily worked by a child, and accomplishes in two hours what is now a hard day's work.

The Washer is made to boil, and is kept boiling, by a gas-burner placed underneath; thus it is continually throwing off a large quantity of steam, which forces its way through all parts of the clothes, and in so doing carries away every particle of dirt, and leaves the clothes spotlessly clean. The clothes only need steeping in water for a few hours, or overnight, then wring them out, soap the dirty parts, and when the water in the machine boils, put them in the cylinder, and turn the handle for ten minutes, very slowly, then take them out and rinse thoroughly, blue and wring out, and they are ready for drying.

A Novel Cycle Boat.
(Zimer's Patent.)

There has been a deal of talk about new cycle boats and launches; here we are able to give our readers some definite information about one of them.

The illustration shows a schematic front view of a "Zimer boat," with the hull and floats in cross section, and an operator working a screw propeller and automatically manipulating the so-called "Zimer floats." The centres of the movable frames which carry the two floats Er and Br run parallel to each other, and a stiff cross bar R, with a knuckle joint at each end, establishes a connection between the floats in such a manner that they move freely but in opposite directions to each other; *i.e.*, if float Er is moved upward float Br is pushed downward by the cross bar. A handle bar on a vertical post furnishes an easy and suitable rest for the operator's hands, and is in direct communication with the cross bar and the two floats.

A slight turn of the handle bar causes the floats to assume a swinging motion, and on the occupant feeling any tendency to lose his balance, he will instinctively turn the handle in the direction of his fall, thereby dipping the float on that side into the water and arresting, with great ease, the rolling motion in its first stage, so that, if riding on the side of a wave, only a slight pressure with his hands will be needed to counteract the one-sided pressure of the water against the boat. It is nothing more than a balancing action, with the important peculiarity that the boat must always return to the vertical line. Any slight movement beyond the balancing point will be followed by a corresponding repulsion to the opposite side. The shaded part of the accompanying illustration represents the boat riding on a wave (R). The dotted lines H indicate the position of the operator's arms and the floats when the boat rides on horizontal water, or on the top of a wave. The dotted lines L show the position of the operator's arms and the floats when the boat is riding on the other side of the wave. There is no skill required in operating the floats, the action is, as we have already said, instinctive. Suppose, for instance, the boat is in its normal position on horizontal water, as indicated by the dotted line H, and a wave (R) is approaching: the swell will reach float Br first and raise it, the operator's hands resting yieldingly on the handle bar, and as it rises it pushes down the opposite float Er as far as the water will permit, so that when the swell reaches the boat the floats have already assumed their proper balancing position. As the boat passes over the top of the wave, the water rising under float Er pushes it upwards and float Br accordingly descends, and so the boat retains its equilibrium. By this means one man can balance it sufficiently for a crew of three or four, and it is evident that, besides being free from the usual rolling motion, it is almost impossible for the boat to capsize. The propelling mechanism consists simply of the pedal crank or treadle, such as is used on the ordinary road cycle, from which the power can be transferred, either by means of bevel wheels to a shaft with a screw propeller, or by means of rods or chains to a stern paddle wheel. For general use, and particularly in rough water, the screw propeller is preferable; but when extremely light draught is necessary the stern paddle wheel is found satisfactory. When the boat has only one occupant he can operate the cycle and steer by means of a separate lever, or by using, when necessary, one hand for steering and one for balancing the floats. If, however, there are two or more occupants aboard, one of them can manipulate a steering handle, besides performing a share of the propulsion.

The balancing arrangement already described admits of the boat being built on very fine lines and yet keeping steady and upright in the water without the need of ballast or a centre board. It is claimed that no rowing boat can keep pace with it in rough

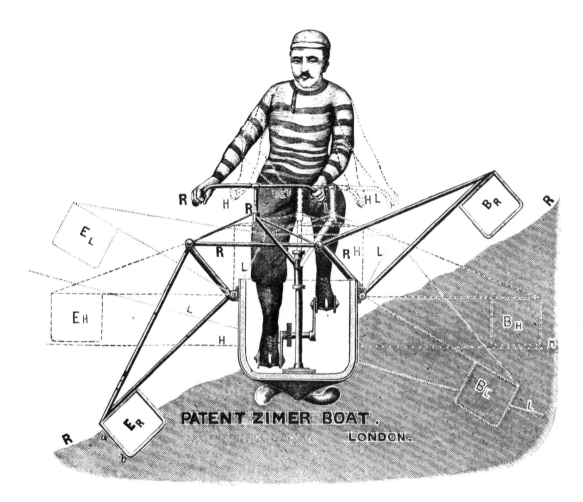

PATENT ZIMER BOAT.
LONDON.

water, and that, since it can be safely constructed to suit speed exclusively, after further experiments it will be able to beat the faster sculler on smooth water in long distance matches.

Other advantages claimed for the boat need only passing notice. It is said it can be manipulated without any special skill, such as rowing or sailing demands, and the legs being more powerful than the arms, and not so soon tired, long distances can be accomplished without the hard work and fatigue of rowing. It also does away with the need for crouching in uncomfortable attitudes on low seats and rowing with the back to the bow of the boat, so that the oarsman cannot see where he is going, and enables the operator to occupy a position from which he can look ahead and not only enjoy a good view of the scenery without stopping, but also see everything that may happen to be in his way and so avoid collisions.

The most popular use of this boat will probably be for pleasure purposes at the sea side, where so many holiday seekers are deterred from boating by reason of the great strength and skill required to pull against the sea and the little enjoyment to be obtained from being rowed about by a professional hand. It can, however, it is claimed, be adapted to numerous other purposes. Its steadiness and the facility with which it is said it can be manoeuvred should make it especially useful as a lifeboat; its swiftness and the fine lines upon which it can be built should recommend it as a racing boat; and for carrying despatches, for harbour, police and revenue service its swiftness again; and the ease and safety with which it can be steered on a busy river, are said to be commendable features. In short, it is claimed that for all matters, of either business or pleasure, these boats will be found equally well adapted.

The invention is said to have been adequately tested on rough water in windy weather and to have stood the severest tests with ease.

The "Cyclone" Home Trainer.

A decidedly useful invention—one which we are quite convinced will be largely adopted in this country—has been made by Mr. L. F. Guignard, of Lausanne, Switzerland. It is the "Cyclone" realistic home trainer, which has been patented in different countries, and which imparts the great pleasure of riding a bicycle on an ordinary road, without leaving the room or place in which the apparatus is arranged. The accompanying illustration shows the principle of the apparatus.

The "Cyclone" is the nearest approach to road riding. It constitutes a vindication of the sound principle of the bicycle. It is an open secret in mechanics that the balance is not maintained by centrifugal force or abstract speed. The popular conception is erroneous. The true principle of the single track machine, whether "ordinary" or safety, is the turning of the wheel towards the falling side. M. Guignard has faithfully applied this principle. The realism is also noticeable in the fact that riding out of the middle of the roller increases the distance travelled, as does

wobbling on a road or riding too "wide" on a track.

The "Cyclone" home trainer consists of a bicycle placed on a stationary platform. Underneath this are arranged two drums which are geared, and revolve together, and which have concave faces. The upper surfaces of the drums pass through the top of the platform, and, as will be seen from the illustration, it is upon these drums that the operator rides, or rather they form the road for the rider, and are revolved by him in riding the bicycle.

Many more gentlemen would learn to ride bicycles but they object to the first lessons on a public road. These may, however, now be more efficiently taken on the "Cyclone". A Lausanne paper in a recent article upon this invention spoke of it as a "truly discreet trainer" for young lady bicyclists. Any person can more readily acquire the delightful art of equitation when the machine is working stationary. The balancing action is present, and the dangerous motion absent. The only risk has been the speed which prevents learners from easily regaining their feet, and causes violence of contact with the ground in falling. Much use will, doubtless be made of the "Cyclone"

MACHINE FOR DISCREET HOME TRAINING.

in theatres, &c., and especially upon the Continent. The interest of indoor riding is greatly increased upon the "Cyclone" by riders seeing before them upon dials the exact distance traversed, and every variation of speed. The maximum pace is also recorded by a special needle which is pushed forward and left something like a self-registering thermometer.

The "Cyclone" home trainer will be found very useful in club rooms, cycle depots, gymnasia, theatres, in small rooms at home, &c. It is fitted with a brake which can be applied as desired, and thus represent the ascent of a hill. As already mentioned, a special apparatus records the distance, whilst a tachometer indicates the variations of speed. The "Cyclone" can be well recommended. It will, we are sure, be highly appreciated.

A New Mechanical Horse.

The Western Mechanical Company, of Exeter, offers the public a novel appliance in the shape of a Portable Exercising Machine or Mechanical Horse, the invention of Mr. Geo. H. Ellis, of Fonthill, St. Leonards. Its object is to provide healthful and pleasurable exercise at home, in addition to the function of an ordinary seat or rather music stool. As shown in our engraving, it is in a pedestal form, similar to a music stool, and can be regulated to suit individuals of different heights. The adjustable seat is mounted upon a steel spiral spring, within which are arranged padded buffers, fitted in a manner to regulate the distance between them ; or the buffers can be tightened together, by which means the stool is rendered firm for ordinary use. When they are separated, the body is carried upon the spring, and can be vibrated up and down, giving a gentle bumping action or concussion upon the buffers at each depression. The movement obtained by the apparatus resembles the action of a horse. The body can be swayed about in any direction, and full play given to the arms and chest. It is claimed for this appliance that it does not keep the body in one position like other seats, with the attendant evil of inducing a sluggish action of the liver, but provides a wholesome and natural means of

A NOVEL APPLIANCE.

stimulating and helping the digestive and other organs to perform their functions, without that constant resort to medicine which so many persons of sedentary habits employ. The apparatus may be used out of doors, on a lawn, under a verandah, or in any other convenient place, and by that means fresh air with healthful exercise can be obtained. When converted into a rigid position by bringing the buffers into contact, the article can be used as a music stool, or as an ordinary desk or house seat. It is recommended for long sea voyages, during which sufficient exercise is difficult to obtain. It can also be conveniently carried about when travelling as personal luggage, and should be a useful and popular addition to the gymnasium ; also for hydropathic establishments, schools, hotels, boarding houses, clubs, and for public institutions generally.

FIG. 1.

FIG. 2.

The "Sultan" Domestic Turkish Bath.

Our engravings show the "Sultan" patent domestic Turkish bath, which offers very considerable advantages to those desirous to take Turkish baths at home. It is often very inconvenient to go to the usual Turkish bath establishments, in vogue in most large towns, and further, many persons feel that these places lack the privacy and comfort of a bath at home. The "Sultan" Turkish bath completely overcomes these drawbacks as the bather can have it in his bedroom, and the arrangements for working it are so simple that it requires no assistance, and the merest novice can be completely instructed in the use of it in a few minutes. As will be seen from our engraving Fig. 1, which shows the bath open, it consists of a rectangular cabinet, very stylishly got up, about 5ft. long, by 2ft. wide, with a door in front, and a hinged lid at the top. The inside has a very comfortably shaped couch made of hardwood laths, with spaces between so as to allow of the free circulation of the hot air; underneath the couch is a hot-air chamber, which, in this patent bath, is heated from the outside by means of a curved and movable flue, and under which is placed either a gas, petroleum, or spirit-lamp or stove, as may be most convenient. All other baths of this type have hitherto had the lamp or stove inside, with the result of great heat or scorching in one part of the body, and insufficient heat elsewhere, but in the "Sultan" bath this is entirely obviated by the hot-air chamber just referred to, which most comfortably diffuses an even heat throughout.

When the bather has seated himself upon the couch he can then close the door and lower the lid, as shown in Fig. 2, which is shaped to fit the neck, and has also two circular apertures with sliding doors for regulating the heat, and also for passing the arms through, so as to read or smoke, &c., while enjoying the pleasures of the Turkish bath. There is also a conveniently-shaped and adjustable head-rest which makes it additionally comfortable.

To further increase the health giving properties of this bath, it can be fitted with a complete electric battery, entirely under the control of the bather, and capable of giving a direct current to any and every part of the body; it can also be very readily converted into a Russian vapour bath for a few shillings extra.

Trouser Stretchers.

There are numerous trouser stretchers already on the market, but we give an illustration of a self-acting stretcher which is the invention of Mr. Joseph Bedford, Haymarket Chambers, Sheffield. The method of working the stretcher is as follows: Fold the trousers and place them in one of the positions as shown in illustrations, the springs are then compressed by the hand until the ends of the springs are able to be inserted three or four inches up the leg of the trousers, the spring is then released and will be found to fit firmly in the leg of the trousers. The springs can then be adjusted as wished and any desirable weights can be attached. The trousers should be left in this position until required for wear. The trousers, of course, can be hung in a wardrobe or wherever wished. The trouser stretchers are placed upon the market at a reasonable cost, and there is no doubt that they will have a ready sale.

New Method of Locomotion.

FIG. 1.   FIG. 2.

Our engraving shows a new mode of utilising the principle of stilts for locomotion, which has been patented by Mr. Samuel Davies. The action of propelling is that of skating on ice, and any forward figure that can be done on ice can be accomplished with ease by these machines. Each wheel is independent of the other, and backward travel is prevented by a mechanical action. An idea how to learn to ride them is given in the engraving. The balance is the first movement to be learnt. By pressing the thumbs on the brakes the wheels become fixed, by which means the learner can walk on them the same as on stilts. When the balance of walking is acquired, the learner may gradually let go the brake on one side for the wheel to move a little forward, then fix the brake on the wheel advanced, let go the brake on the opposite side, and advance that wheel a little in front of the other, always, however, taking care to brake the front wheel before advancing the hind one. By this means the action of the wheels moving under you is acquired, but it is advisable to go slowly to work at first. When the balance is lost it is best to jump off the machines and commence again, as the rider is not fixed in any way on the machine. The treadles are free to move with the toe or heel so that the control of the machines when in motion can be retarded by pressing the toe down, or the reverse, by pressing the heel down. The handles have the same effect: by throwing forward they retard the wheels, by placing them backward they assist the progress.

A Novel Portable Couch, or Sleeping Rest, for Travellers.

The invention illustrated in the accompanying views is designed to provide convenient means of making up a couch, seat, leg rest or table for the use of travellers in railway and other vehicles, or on board ship, and for other like purposes.

A folding framework of wood, metal or other suitable material is provided, which may be covered with canvas, netting or any other flexible material, as indicated. The whole, when extended, may be arranged to rest on the seats of a railway or other carriage, or upon chairs or other convenient supports, as will be understood upon reference to the annexed illustrations. In construction, the device may be made of several parts sliding one within the other after the manner of a telescope, or it may be formed of lattice work or of separate parts hinged to each other, so as to fold or roll up. In the last type of construction the contrivance is equipped with a number of straps or buttons attached thereto so that they may be turned across the hinged joints to maintain the whole in a rigid compact form.

It will be understood that this device when constructed according to any of the before-mentioned plans may be readily folded up into a small compass when not required for use, which renders the appliance capable of being stowed away.

FIG. 1.

Fig. 1 of our illustrations represents the interior of a railway carriage with the passengers, under ordinary circumstances, endeavouring to obtain rest with the primitive and unsatisfactory means at their disposal. Fig. 2 shows a similar view, but with the occupants of the carriage comfortably reclining or reposing upon Mr. Newton-Peake's patent portable sleeping rests as before described. An open plan and a view of the contrivance folded up are shown in Fig. 3, whilst Fig. 4 represent a back

FRONT VIEW   FOLDED   FIG. 3.

BACK VIEW   FIG. 4.   PACKED

view of the open rest and the device packed up neatly in a travelling rug respectively. The sleeping rest will be found to give ease and comfort to the tired traveller, and as the apparatus complete weighs only 5lb., its lightness and portability appear all that can be desired. In the underside view, two small transverse "stretcher bars" are shown attached for the purpose of maintaining the canvas, or fabric, taut, which may project for a convenient length beyond the longitudinal frame, bars or laths, as shown in Fig. 4. The length of the rest when folded is 36in. and the breadth 3½in.

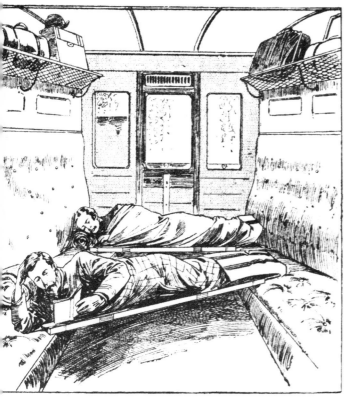

FIG. 2.

Sitting while Asleep.

A very ingenious invention comes from Germany, which enables the user to rest as comfortably and safely as if lying on a bed, as it provides a rest for head, neck, back, and elbow at the same time. The invention will be readily understood from our illustration,

and it is claimed for it that the appliance is especially useful in the case of travelling for long distances by rail. It is also easily packed away in a small parcel, which can be carried in the pocket. It is claimed that by using the invention the traveller will hardly feel the shaking of the railway carriage, while he can at any time by means of a single turn change his position as he likes by leaning to the right or left, or sitting straight, but in any case there is a firm support for his head. The appliance is especially useful for asthmatic persons and those with chest complaints, and we understand that eminent German physicians speak most highly of the invention. The appliance is also used for sitting upright in bed, in which case it is fastened to a hook fixed in the ceiling. As the price is moderate, this excellent invention should meet with considerable support in this country. The inventor is Mr. P. Knüppelholz.

The Electric Lighting of Railway Carriages.

The question of providing a greater amount of light in railway carriages has for years past occupied the attention of railway managers, but a pardonable solicitude for the company's purse has prevented most of them from effecting such an improvement as would silence the constantly repeated complaints of their passengers. It is true that on many lines gas has superseded oil, but the increased illumination is scarcely more than was required to enable the passenger to enter and leave the train in safety, while the possibility of reading in comfort is still out of the question.

An increased roof light, moveover, is scarcely what is wanted by the travelling public, as the passenger who desires to read requires a separate light near his book. To the Metropolitan District Railway Company belongs the credit of being the first to adopt a system of electric lighting which supplies this want, and which, while it does not call for any expenditure of capital by the railway company, gives the travelling public just what is wanted, and at the same time admits of a satisfactory profit being made by the company undertaking the supply. This want is met by the railway electric reading lamp, which is a machine invented by Mr. Tourtel for retailing electric light to passengers by pennyworths.

A few of these lamps have been experimentally in use on the Metropolitan District Railway for the past two years, and the results of the trials have been so satisfactory that a contract has been made by that company with the Railway Electric Reading Lamp Company to instal up to 10,000 reading lamps in the carriages of the District Railway. It is not intended at present to displace the gas lamps in use in the roofs of carriages, but to provide a separate light for passengers desiring to read.

The mechanism by which this will be accomplished is exceedingly simple, and is contained in a box 5in. by 3in. Upon introducing a penny into the slot at the top of the machine and subsequently pressing a knob, an electric light is obtained which burns for about half an hour, at the end of which time it is automatically extinguished, but can be relighted by the insertion of another penny.

TOURTEL'S APPLIANCE FOR PROVIDING LIGHT BY THE PENNYWORTH.

The light, which is of about 3-candle power in strength, is concentrated by a shaded reflector, which may be turned within certain limits so that the light may be directed to suit the position of the passenger.

One of the most remarkable of the many unprecedented features of the instrument is its honesty, as it is so arranged that in case of a failure in the supply of electricity, the machine automatically returns the coin to the operator. Another noticeable feature is that, should the lock of the apparatus be tampered with, a bell is automatically set ringing in the guard's van.

It is proposed to place these lamps under the hat rails in railway carriages, so that the passengers seated in the corners, and thus furthest from the roof light, may be able to provide themselves with additional illumination. The whole of the lamps placed in one carriage are supplied with the electric current from an accumulator placed under one of the seats of the carriage, which is thus entirely self-contained, and capable of being detached from the train without the light being affected. The accumulators can be easily changed, and will be replenished at charging stations near the terminus.

The lamps are worked at the low pressure of 12 volts, and consume about three-fourths of an ampere. With this exceptionally low pressure, there is absolutely no danger of any kind.

The accumulator battery has been specially designed for the work, and consists of six cells, coupled in series, having a capacity of 72 ampere hours. The cells are enclosed in a strong wooden case, which is provided with rollers and handles for convenience in removal, and duplicate sets of accumulators are provided for each train.

An accumulator battery charging station is to be erected at Mill Hill Park, on the District Railway, and sidings are to be made in the station for the reception of a number of trucks on which the accumulators will be charged and removed. Early in the morning trucks containing the charged accumulators will be forwarded from the charging station to the various distributing termini, where they will be changed as required during the day; and then at night the trucks will return with the exhausted batteries. As each battery will contain a sufficient supply of electricity to suffice for over two days, it will only require to be changed every alternate day, and thus only half of the duplicate batteries will be charged daily.

An authority upon electrical matters as affecting railway companies—Mr. C. E. Spagnoletti, past-president of the Institution of Electrical Engineers and electrician to the Great Western Railway Company—has stated that the advantages to railway companies will induce them to adopt this system generally. Mr. Tourtel's system also promises a solution of the question of railway carriage roof lighting.

The supply of light under this system promises to prove very remunerative to the company, and at the same time the price charged for the light is within the reach of all railway travellers. The passenger would certainly prefer to spend 2d. for two half hour's electric light, than buy a candle and suffer all its attendant annoyances.

A Simple and Cheap Form of Portable Illumination.

Messrs. von Hoevel, of East India Avenue, is introducing the cheap and simple little lamp which forms the subject of the annexed illustration. The body of the lamp holder is composed of stamped metal, provided with convenient means of suspension and for clipping the same on to a book or the like. The lamp itself consists of a little metal receptacle containing the wick and oil, or fatty matter, which may be fitted into the holder mentioned. A number of the charged receptacles are sold in a box with each holder. The novelty is worthy of notice by the travelling public, who frequently desire a cheap and more efficient means of illumination than that often provided by the railway companies.

CHAPTER III.

Stringent regulations are currently being laid down in relation to the means of egress and escape from buildings where there are many occupants and working people—as boarding houses, hotels, apartment houses, tenement houses, factories, &c.—all of which are to be kept under the most rigid inspection by officers appointed for the purpose, and these laws, if thoroughly enforced, will go far to prevent in the future the fearful holocausts of the past. In this chapter we review a number of the latest devices which our inventors have brought out in compliance with these laws, as well as others designed to fight fire as opposed to escaping from it.

Anidjah's Portable Household Fire Escape.

This fire escape is very simple in construction, and it may be said of it that a child can use it with the greatest ease and without any risk whatever. It consists of a neat wooden box measuring only 9in. high by 30in. long and 18in. wide, which, when required to be used is moved against the wall of the room immediately under the window. The lid is then raised and placed against the wall, and the front of the box falls flat on the floor, disclosing a strongly-made canvas shoot or pocket open at both ends, flanked on each side with ropes long enough to reach across wide thoroughfares. Attached to these ropes are two large plaited rings. The side ropes are called guide lines, and when the window is opened the plaited rings are thrown out into the

street to bystanders, or in their absence the inmates can be lowered by simply passing their arms through the rings leaving their hands free to push themselves away from any projection obstructing their descent. These rings can either be held or can be placed on the tops of railings of opposite houses, or tied to posts or any other convenient object.

The shoot is then thrown from the window and travels along the guide lines by means of iron rings securely fastened to each side the whole length of the canvas. The under part of the upper end of the shoot is fixed to the box, the upper part forming a flap which covers the space of the open window and is kept in that position by two hooks, one on each side of the window

FIG. 1.

FIG. 2.

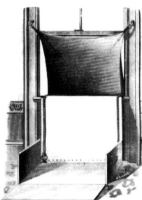

FIG. 3.

ANIDJAH'S PORTABLE FIRE ESCAPE: A VIEW OF THE APPLIANCE IN USE.

27

frame. The mouth of the shoot being thus formed it is absolutely impossible for any person to fall or to drop another into the street by accident.

If anyone, through nervousness or faintness or any other cause, were to stumble or fall on reaching the window, or should an infant be dropped from their arms they could only fall into the shoot and would descend, even in an insensible condition, in perfect safety. The whole of the apparatus is rendered uninflammable by the application of chemicals.

Should it happen that no persons were in the street at the time of the fire, which is very unlikely when such a calamity occurs, anyone can descend in quite a perpendicular position by merely pressing their elbows and knees against the sides of the shoot while descending. Should there be balconies or areas or any other projections, the inmates can lower themselves by means of the guide lines.

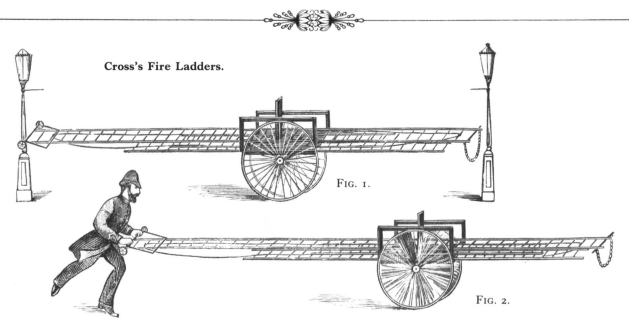

Cross's Fire Ladders.

Fig. 1.

Fig. 2.

One of the latest inventions in the department of fire escapes is the patent fire ladder which has been brought out by Mr. John Cross, of London, and which is intended as an auxiliary to the existing methods employed by the Metropolitan Fire Brigade. The apparatus consists of three light ladders of varying lengths, reaching to a height of 45ft., together with ropes and pulleys, the whole weighing only 3cwt., being easily portable by means of a small two-wheeled car. Two persons are required to put the apparatus in position, a task occupying only a few minutes. The chief recommendation of the invention appears to lie in its portability, and simplicity and cheapness. By an ingenious manipulation of an endless rope, worked on a pulley attached to the top of the ladder, rescues can be simultaneously effected from each storey of the house. The system also comprises escape bags and belts, which are carried by the rescuers, and quickly and easily adjusted. It is suggested that Mr. Cross's invention, by being placed in the public thoroughfares at convenient distances, would be found of real usefulness in the absence or pending the arrival of the fire brigade apparatus. The ladders and accessories would, of course, be the property of the governing body of the district, and would be kept under proper care for the public benefit. Our engraving, Fig. 1, shows the apparatus ready for running to a fire. Fig. 2 shows it going to a fire, and Fig. 3 in actual use. It will be seen that these ladders are always ready for immediate use, they require no minding, and can be used by anyone of moderate strength, without previous drill or training, while two men can raise the longest ladder.

FIG. 3.

These ladders can, moreover, be used in a great many places (especially suburban) where the usual street fire escape would be unavailable, will go through narrow door-ways.

A number of persons, even though insensible, invalids and children may be got from danger to safety with comfort and rapidity by the police and others. We may state that repeated tests have been successfully carried out in various parts of the

Metropolis, and we congratulate Mr. Cross upon having devised a very efficient and useful apparatus, which seems likely to play an important part in the saving of life from the dangers of fire.

"Bucket Nest" Fire Extinguisher.

Messrs. Messer and Thorpe are the patentees and manufacturers of a new arrangement in fire extinguishers which combines immense simplicity with undeniable efficiency. The inconvenience and unsightliness of keeping a number of filled buckets suspended in a room or hall prevents many from availing themselves of these "first aids" to extinguishing a conflagration. The defects above mentioned, however, are entirely avoided by the apparatus of which we give an illustration on the next page.

As will readily be perceived this consists of a system of telescoping the buckets together beneath the water in a covered receptacle, whereby 10, 20, 30, or more buckets can be stored in a comparatively small space; and the inconvenience and unsightliness of exposed buckets, together with the repeated labour and attention required in refilling, are entirely avoided. The latter advantage is very important when it is considered how often replenishing even in public institutions is neglected, and discovered only when the consequences are irretrievable. The buckets, by the ends of their handles, are guided into, and secured in their places. They are released by the simple action of raising their handles when lifting them out. By this arrangement the buckets on being withdrawn pick up their water and come out full, ready for immediate use, and can be taken out in very rapid succession, as it is impossible for a bucket, while being removed, to lift by suction those beneath it, or to become in any way jammed. The water, treated with a little disinfectant (such as Condy's fluid or permanganate of potash) and contained in a closed vessel, under seal (only to be broken in the event of fire), is perfectly protected from contamination, evaporation, and misuse.

The bucket fire extinguisher has no complicated parts, taps, or other fittings, liable to get out of order through long disuse.

Messer and Thorpe's Bucket Nest

It is very compact, occupying so small a space that it can be placed almost anywhere, and may be made—where its normal appearance might be considered inadmissible—in the form of a pedestal, sideboard, settee, seat, &c., *en suite* with the furniture.

Even at sea where the water is all-surrounding a "nest" of the buckets will usefully replace the existing ranges. In houses, factories, mills, museums, and all public buildings whatsoever this device is incomparably superior to anything now in use; and the only wonder is that it was not thought of before. The absence of any mechanism and the fact that it renders a large quantity of water at once available, are high recommendations, and it will doubtless command a large sale.

H. Hargreaves' Patent Window Fire Escape Apparatus.

It will be immediately apparent that much labour and thought has been bestowed upon this apparatus in order to bring it to its present simplicity and safety. The whole of the test and trials carried out during its development were conducted from a second floor window. The points aimed at in its construction

were rapidity of action, simplicity, absolute safety and absence of unsightliness, and a study of the apparatus shows how complete is the success attained. Fixed under the window on the top floor, or on a lower floor if preferred, the apparatus is covered by a dressing table top which entirely hides it from view. It is thus always ready for use. On an alarm of fire the table top is removed and the apparatus thrown out of the

H. HARGREAVES' APPARATUS: DETAIL.

window ready for action. That operation is exceedingly simple, and if done with alacrity takes about eight seconds. If using a large size apparatus two adults or three children can be lowered at one time. Two persons having passed off the platform, stretching across the apparatus into the bag, the person at once lowers them to the ground by means of ropes attached to the bag and coiled upon runners fixed upon the shaft under the window. A hand brake governed by a spring and hand lever gives the operation perfect control in thus lowering the escaping persons. Having lowered the first lot he winds up the bag by

means of a winch handle, and passing others into the bag lowers them also. From a house of average height from twelve to fifteen people can be got out in four minutes, with ease and perfect safety. Having lowered all the occupants the operator proceeds to lower himself or herself. To do this he attaches two hooks fixed to lines on rollers in the bag, to the iron frame of the apparatus. Having thus attached the hooks he takes off the hand lever brake by pressing it down and inserting a wedge, and then passes into the bag. Kneeling down and holding the lines which are by the hooks attached to the iron frame, he presses with his knee a lever which releases the rollers at the bottom of the bag, and thus gradually descends to the ground. Perfect control it will thus be seen is afforded to the last person in effecting his own escape, just as hitherto he had had perfect control in lowering others. Young people of either sex, after a little instruction, are as capable of using the apparatus as grown people. Perfect independence of outside aid is afforded by this invention, and from its construction, no matter at what height it is fixed, there is no feeling of uneasiness in descending. Invalids and weak persons can be easily passed into the bag.

Avoiding the Danger of Gas Explosion.

The new "Bo Peep" gas extinguisher is an ingenious arrangement by which the gas is never turned completely out although invisible, thus avoiding the explosion by the taps being left on when the gas has been turned off at the meter. It is of advantage to persons of a nervous disposition, as by pulling a small cord a light can be immediately obtained and again by pulling another put practically (not really) out. These cords can be arranged along pullies to side of the bed or to any part of the house as may be desired.

32   FIRST DESPATCH OF MAILBAGS THROUGH THE PNEUMATIC TUBE FROM EVERSHOLT STREET TO EUSTON STATION.

CHAPTER IV.

NEW FORMS OF TRANSPORT.

A TYPICAL example of the pioneering spirit of our inventors and engineers, particularly in the field of transport, is the development of the pneumatic despatch mail service. A Company, registered for the establishment of lines of pneumatic tube for the more speedy and convenient circulation of despatches and parcels, has been the subject of an Act of Parliament, which has now received the Royal Assent, is empowered to open streets and lay down tubes for the purpose.

The directors, having satisfied themselves and the shareholders of the complete mechanical success of the company's system of transmission, by experiments upon a short line of tube at Battersea, and of its economy and peculiar applicability to the purpose in view, determined on laying down a permanent tube of thirty inches gauge between the Euston-station and North-western District Post-office, Eversholt-street. This tube, with the station, machinery and appliances, is completed, and is found to work most efficiently. The length of the tube is not considerable, reaching a distance of only a third of a mile. The transmission of the first batch of mailbags recently took place. Several of the principal officials from the Post Office were present during a part of the operations.

The whole of the works were in the most admirable order, and on the arrival of the first mail-train at 9.45 a.m., the mailbags, thirty-five in number, were placed in the cars by 9.47. The long chamber was then exhausted and the train containing the first mails ever despatched by the agency of the atmosphere, were blown through the tube to the station at Eversholt-street reaching their destination at 9.48. The success of the experiment has been so decided that the company will commence the Holborn extension at once, and intend proceeding as rapidly as possible with the main work and all its ramifications.

It may be of some interest to our readers to review the developments which preceded the Eversholt-street experiment. The novel and enterprising suggestion for the speedy conveyance of letters and parcels by means of a pneumatic tube was embodied in the creation of the Pneumatic Despatch Company, only a few years ago. It was decided to proceed with preliminary tests of the principle, which, essentially, consisted of creating a vacuum before the vehicles, the tube being, in fact, the cylinder, and the carriages the piston.

A piece of ground adjoining the Victoria Railway-bridge at Battersea, and belonging to the Vauxhall Waterworks Company and London and Brighton Company was selected for testing the project. Upwards of a quarter of a mile of the tubing was laid down; various irregular curves and gradients were introduced to show that hills and valleys would not prevent the effective working of the system. The apparatus worked very well. With an exhaustion varying from 7 in. to 11 in. of water, or from 4 oz. to 6 oz. per square inch, the speed is about twenty-five miles an hour. The tube through which the despatch-trucks are drawn is not circular in form, but of a section resembling that of an ordinary railway tunnel; the internal height being 2 ft. 9 in., the width at the springing of the arch (the top being semicircular) 2 ft. 6 in., and at the springing of the invert (for the tube has a segmental bottom) 2 ft. 4 in.

The tube is of cast iron, in 9 ft. lengths, each weighing about one ton, and fitted into each other with an ordinary socket-joint, packed with lead. Within the tube, and at the lower angles on either side, are cast raised ledges, 2 in. wide on the top, and 1 in. high, answering the purpose of rails for the wheels of the despatch-trucks to run upon. The latter are made of a framing 7 or 8 ft. long, inclosed in sheet iron, and having four flanged wheels, 20 in. diameter each. The whole truck is so made that its external form, in cross section, conforms to that of the tube, although it does not fit it closely, an intervening space of an inch or so being left all around. Some light india-rubber flanges or rings are applied at each end of the truck, but even these do not actually fit the inner surface of the tube, a slight "windage" being left around the whole truck. There is therefore no friction beyond that of the wheels; and the leakage of air, under a pressure of four or five ounces per square inch, amounts to but little.

The air is exhausted near one end of the tube by means of an exhausting apparatus, from which the air is discharged by centrifugal force. Some idea of this apparatus, which is very simple, may be formed by comparing it to an ordinary exhausting-fan. (See the engraving on page 6.)

It is the intention of the Company, now that they have obtained Parliamentary powers for opening the streets to lay down their tubes, to establish lines between a large number of points throughout the metropolis, ultimately extending the system so as to connect the Railways with all public offices.

In the Battersea experiments, one trip was made in sixty seconds, and a second in fifty-five seconds, the distance being a quarter of a mile. Two gentlemen occupied the carriages during the first trip. They lay on their backs, on mattresses, with horsecloths for coverings, and appeared to be perfectly satisfied with their journey. It is calculated that the carriages will eventually move through the tubes at the rate of from thirty to forty miles an hour.

Other Atmospheric Railways.

The subject of pneumatic traction has been under active consideration and research for some years. It was as early as 1844 that it was announced that trials on the Atmospheric Railway between Kingstown and Dalkey, in Ireland, had been successfully

completed. This was an experimental stretch of railway, laid down originally for the purpose of conveying granite from the quarries of Dalkey for the construction of the magnificent harbour of Kingstown, in which the train was moved from point to point by atmospheric pressure. Here, however, instead of the whole body of the vehicles being subjected to the influence of the vacuum, a pipe of about twelve inches diameter was laid down between the rails and a travelling piston, affixed to the underside of the carriages, forced along the pipe by air pressure. A rod or plate of iron, connecting the piston with the carriages, traversed a slit on top of the pipe. The great difficulty to be overcome was to cover this slit with a substance which would be air-tight, and yet would permit the connecting-rod to pass without suffering much obstruction. The problem was solved by providing a continuous valve, the whole length of the pipe, constructed of leather riveted between two iron plates. As the travelling piston is forced along the pipe the valve successively opens and closes up

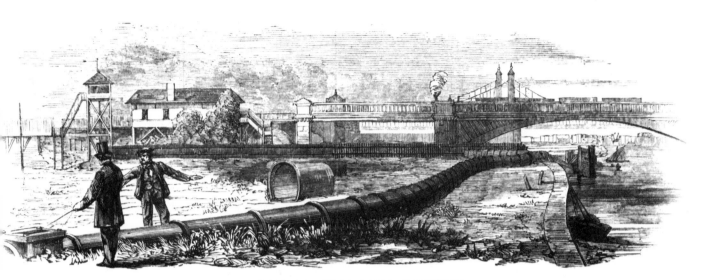

THE EXPERIMENTAL DESPATCH TUBE AT BATTERSEA.

again when the piston has passed. The occlusion of the valve is made completely air-tight by means of a composition of wax and tallow, which, when cool, solidifies, and when warmed, melts. A copper heater, filled with burning charcoal, passes over the composition and melts it, thus leaving the valve as air-tight as before, and ready for the next train.

As compared with rail traction by rope, to which plan it bears the nearest resemblance, the system is immeasurably more effective. In the following year, 1845, Dr. Hewlett, in a paper read before the Society of Arts, described an improved version of the system in which, by an ingenious system of valves and valve-releases, the disadvantages of a continuous valve, with its inherently high risk of leakage, and consequent requirement of greater air pressure, were obviated. Whereas the Dalkey-Kingstown system required for the maintaining of adequate pressure a pumping station at intervals of three miles along the track, the new system needed only one at every ten miles. The

discontinuous valve also did away with the need for the charcoal heater. As with other methods of atmospheric propulsion, Dr. Hewlett reminded his audience, there was the advantage of the avoidance of collision (as it is impossible that there can be two *vacua* in opposite directions on the same line); the avoidance of smoke, steam, and falling flakes of fire; the rapidity of progress— and the extraordinary saving of expense.

In the same year there were demonstrations of the working of the Croydon Atmospheric Railway, an experimental stretch laid down at Wormholt Scrubbs over a length of three-quarters of a mile. Subsequently, five miles of the track were completed, and in trial runs it was reported that a train of twelve carriages reached speeds of up to seventy five miles per hour. This system was also adopted by the South Devon Railway and by the Paris and St. Germain Companies.

The mechanical success of the atmospheric system is certain. It appears only a matter of time before it becomes general.

35

THE AIR-OPERATED PASSENGER RAILWAY AT THE CRYSTAL PALACE, SYDENHAM.

There is perfect command of speed and greater facility of arresting motion. There is less oscillation and less vibration. There is no necessity for a *long train* of carriages, for, on the Compressed Air System it is not necessary to link even two carriages together; each may take its own impulse from the continuous reservoir independently, and they may follow on the same line at any intervals of distance.

The Pneumatic Railway at Crystal Palace

We give an Illustration of the Pneumatic Railway, invented by Mr. T. W. Rammell, C.E., a working model of which has been exhibited in the grounds of the Crystal Palace. Instead of drawing its traction power from a tube in the track, this system operates on what we may call "the Eversholt-street Principle."

It extends from the Sydenham entrance to the armoury, near the Penge gate, a distance of nearly six hundred yards. A brickwork tunnel, about 10 ft. high by 9 ft. wide, and capable of admitting the largest carriages used on the Great Western Railway, has been laid with a single line of rails, fitted with opening and closing valves at each extremity, and supplied with all other apparatus for propelling passenger-trains by a strong draught of air behind the train when it travels in one direction, and pumping away the air in front of it when it travels the other way. The motive power is supplied by this contrivance: At the departure-station a large fan-wheel, with an iron disc, concave in surface and 22 ft. in diameter, is made to revolve, by the aid of a small stationary engine, at such speed as may be required, the pressure of the air increasing, of course, according to the rapidity of the revolutions, and thus generating the force necessary to send the heavy carriage up a steeper incline than is to be found upon any existing railway.

The disc gyrates in an iron case resembling that of a huge paddlewheel; and from its broad periphery the particles of air stream off in strong currents. When driving the air into the upper end of the tunnel to propel the down-train fresh quantities rush to the surface of the disc to supply the partial vacuum thus created; and, on the other hand, when the disc is exhausting the air in the tunnel with the view of drawing back the up-train,

the air rushes out in a perfect hurricane from the escape valves of the disc case. When the down journey is to be performed the breaks are taken off the wheels, and the carriage moves by its own momentum into the mouth of the tube, passing in its course over a deep air-well in the floor, covered with an iron grating. Up this opening a gust of wind is sent by the disc, when a valve, formed by a pair of iron doors, hung like lock-gates, immediately closes firmly over the entrance of the tunnel, confining the increasing atmospheric pressure between the valve and the rear of the carriage. The force being thus brought to bear upon the end of the train, the latter, shut up within the tube, glides smoothly along towards its destination, the revolving disc keeping up the motive power until it reaches the steep incline, whence its own momentum again suffices to carry it the rest of the distance.

The return journey, on the contrary, is effected by the aid of the exhausting process. At a given signal a valve is opened, and the disc-wheel set to work in withdrawing the air from the tube. Near the upper end of the tube there is a large aperture, or side-vault, which forms the throat through which the air is exhaled, the iron doors at the upper terminus still being kept shut. In a second or two the train posted at the lower terminus, yielding to the exhausting process going on in its front, and urged by the ordinary pressure of the atmosphere from behind, moves off on its upward journey, and, rapidly ascending the incline, approaches the iron gates, which fly open to receive it, and it emerges at once into daylight.

Instead of a train being used there is one very long, roomy, and comfortable carriage, resembling an elongated omnibus, and capable of accommodating some thirty or thirty-five passengers. Passengers enter this carriage at each end, and the entrances are closed with sliding glass-doors. Fixed behind the carriage, there is a framework of the same form, and nearly the same dimensions, as the sectional area of the tunnel, and attached to the outer edge of this frame is a fringe of bristles forming a thick brush. As the carriage moves along through the tunnel the brush comes into close contact with the arched brickwork, so as to prevent the escape of the air. With this elastic collar round it, the carriage forms a close-fitting piston, against which the propulsive force is directed.

The Cable Subway at Tower-hill.

Not least of the many remarkable developments that the Metropolis has seen in recent years is the subway or tunnel under the Thames from Tower-hill to Tooley-street, Southwark, designed and carried through by Mr. W. H. Barlow, the engineer. The subway consists of a narrow tunnel uniting two vertical shafts, the mouth of one being in Tower-hill and the other in Vine-street, Tooley-street. The tunnel is lined with iron tubing bolted together in short lengths by flanges projecting on the internal surface. The tube is 7 ft. in clear internal diameter, or 6 ft. 8 in. between the flanges, and carries a railway of 2 ft. 6 in. gauge.

On the railway runs an omnibus conveying twelve passengers. The tube is about a quarter of a mile in length, and sinks from both ends toward the centre with a gradient of about 1 in 30. The omnibus is of iron—light, but very strong, and runs upon eight wheels. It is connected with a rope of steel wire by means of a gripe that can be relaxed or tightened at will. At each end of the tunnel the wire runs over a drum worked by a stationary engine. The declivity of the tunnel is such that, when once the omnibus is started, it requires only a small amount of traction, and the momentum acquired during its descent will carry it a long way up the opposite slope. It is said that the strain on the rope never exceeds 2 cwt. The omnibus is provided with brakes, so that its motion is completely under the control of the man in charge. At each end of the tunnel it is received by buffers, or catches, which are connected with very strong springs of vulcanized indiarubber.

The shafts at each end of the tunnel are 60 ft. in depth, and are lined partly with brickwork and partly with iron tubing. Within the shafts are lifts, carrying six passengers at once, and these lifts are raised and lowered by the same engines that work the drums. Each lift has a counterpoise equal to its own weight and to that of three average passengers, representing the maximum of work that will be demanded from the engine, either for raising or lowering.

At the top of each lift is a contrivance by means of which a breakage of the suspending chain would close iron claws upon the lateral guiding-rails and would bring the machine to standstill within the course of a few feet. The ascent of the lifts is checked by means of springs of steel and indiarubber, which the engine employed would not be strong enough to break. The wheel over which the suspending chain runs is also dragged, so to speak, by revolving fans; and too great a rapidity of either ascent or descent seems to be rendered impossible.

The arrangements visible from above are very simple. The upper opening of each shaft is covered by a small square building at the door of which passengers take their tickets, then enter and descend in the lift. On reaching the bottom they find a space of a few feet between the shaft and the buffers fitted up with benches, as a waiting room. When the omnibus arrives, and has discharged its load, those who are waiting step in and start off for the other end. The descent of the shaft occupies twenty-five seconds, and the omnibus journey seventy seconds; so that a passenger may descend into the shaft at Tower-hill and emerge in Vine-street in a minute and three-quarters from the time of his descent.

THE CABLE SUBWAY: THE SCENE IN THE WAITING ROOM AT THE FOOT OF THE TOWER HILL SHAFT.

SIXTY FEET UNDER THE THAMES: INTERIOR OF THE CABLE OMNIBUS.

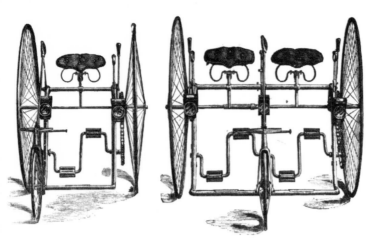

THE DOUBLE-DRIVING CONVERTIBLE.

THE PATENT IMPULSORIA AT NINE-ELMS.

GRASSI'S ARCHIMEDEAN LOCOMOTIVE FOR MOUNTAIN CLIMBING.

THE COMPLETELY CHAINLESS BICYCLE.

IMMISCH'S NEW ELECTRIC DOG-CART.

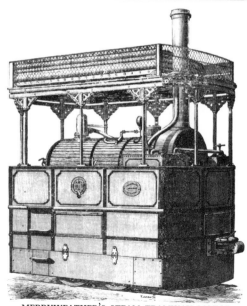

MERRYWEATHER'S STEAM TRAMWAY MOTOR.

A Miscellany of Locomotion.

An ever more varied range of modes of transport becomes available with every year that passes. On this page we give just six recent examples. Firstly we show the Double-Driving Convertible Tricycle, a boon to the outdoor enthusiast. This machine does service either as a "solo" or as a "sociable"; the addition of a simple extension to the framework converts it instantly to allow a companion driver. The front is void of projections which might catch or damage a lady's dress.

The Patent Impulsoria, an ingenious invention introduced from Italy by Signor Masserano, and demonstrated at the Nine-elms terminus of the South-Western Railway, applies animal power to the working of railways. An artificial ground, or platform, is caused by the movement of the horses' hooves to actuate the leading wheels of the locomotive, thus propelling the vehicle. By the movement of a lever the speed of the vehicle may be varied, or it may even be reversed, regardless of the speed, etc., of the animals.

Immisch's Electric Dog-cart was specially constructed for H.M. the Sultan of Turkey and is motivated by 24 small accumulators which contain a charge sufficient to propel the vehicle about five hours at speed of 10 miles per hour.

The new Locomotive Engine for ascending steep gradients on railways, invented by M. Grassi, consists of an application of the Archimedean screw principle.

The Chainless Bicycle is totally different to the two-wheelers as yet brought out, having absolutely no chains or cogs. The pedal motion is quite unique and very easy, being a vertical elliptical motion. The bearings of the driving wheel are fitted in front of the forks, thus rendering a fall over the handles next to impossible.

It may prove that electric traction will constitute a strong competitor to steam, but for the present Messrs. Merryweather's Steam Tramway Motor knows no peer. Its economy, as compared with traction by horse power, is marked. The condensing arrangements are so constructed that steam exhaust is invisible even when ascending hills. All handles—regulator, engines, car brakes, etc.,—necessary for working are in duplicate, so that a man may always drive standing at the forward end.

The Bessemer Saloon Steam-ship.

As our engraving readily conveys, the Bessemer Saloon ship has as its distinctive feature a saloon that maintains an even level, whatever the motion of the vessel. This revolutionary principle is a joint invention of Mr. H. Bessemer, author of notable improvements in the iron and steel manufacture, and Mr. R. J. Reed, C.B., late Chief Constructor of the Navy.

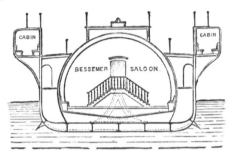

The Bessemer saloon forms by far the finest cabin that has ever been fitted in a ship. Its great size and height enable it to be completely ventilated, unlike the ordinary cabin between decks, which is so unpleasant that ladies and delicate persons endure the worst weather on deck rather than accept shelter in it. But the greatest advantage is that the saloon, being virtually isolated from the hull of the ship, and subject to the action of Mr. Bessemer's hydraulic levelling apparatus, is designed to remain absolutely unaffected, no matter how much the vessel may roll. Mr. Bessemer's approved statement indicates that sea-sickness among passengers is completely eliminated, all sense of pitching and rolling being so small as to be inappreciable. "Mr. Bessemer's hydraulic apparatus is an established certainty, and not a matter of speculation, and it will always insure the floor being kept level."

Our engraving shows the equanimity of the passengers in the saloon area, whereas those on the ordinary deck are subject to the usual discomforts. The illustration was prepared, however, before the ship's first voyage in rough weather, when the device unfortunately proved not entirely satisfactory.

CHAPTER V.

THE INVENTOR AND THE TOILET.

ONE of the most striking signs of the times is a very greatly increased attention to sanitary matters in the ordinary arrangements of the house. The danger to human life is always increased by the accumulation of inhabitants, as in large cities; but, so perfect has become the science of sanitation, that the percentage of deaths, in well-sewered cities, is smaller than that of country districts.

There is, however, no doubt that perfect as the general system of sewage may be, as much importance to the general health is to be attached to the method of connection between the house and the main sewers and to the domestic installation itself.

The number and variety of the devices, now appearing before the public, almost, it would seem, in an embarrassment of choice, is but another illustration of the progress of the present age, and of how science and ingenuity spring ever to the front to meet any acknowledged requirement or want, and this is as it should be; for, with the increasing thousands of our population, we could as ill afford to ignore the responsibility which this subject thrusts upon us, as we could afford wilfully to engulf Her Majesty's navy, with its belongings *en masse*, beneath the waters of the sea. Either act would perhaps be about equally disastrous to human life.

Among the many patent sanitary closets which have been, and are now being, introduced upon the market there is one that deserves special mention. It is the invention of Mr. Heywood, of Manchester, and, from its simplicity of construction, efficiency and durability, ought and no doubt will command the attention of architects and owners of large buildings. It forms a closet and an automatic disinfector combined. The closet, as will be seen from our engraving, is made in the pedestal form and with a trap that effectually prevents the inroad of sewer gas in the house. Behind the basin there are two compartments in which to place the disinfectants; between these is placed the flush pipe, so arranged and perforated on the underside that with each flush sufficient water escapes into each compartment containing the disinfectant. When the flush is over this water forms an afterflush which flows into the basin, thus forming a constant disinfecting fluid always standing therein, so that whenever the closet is used or anything placed therein it is at once thoroughly disinfected and deodorized. The construction of the closet will allow of any disinfectant being used so long as it is of solid form, and to meet this requirement the patentees have produced a special phenol preparation of very great strength, which gives forth an agreeable odour which is not in any way objectionable. This preparation is made in blocks to fit the compartments, and will last, it is stated, from three to four months. Great advantages with these closets are that they are all made of one piece of earthenware, they require no attention on the part of the servants, and cannot possibly get out of order. Several gentlemen interested in sanitary matters have inspected the invention and speak very highly of it, while Mr. Carter Bell, A.R.S., M.F.I.C., the county analyst for Cheshire and the Borough of Salford, states that, in his opinion, "these

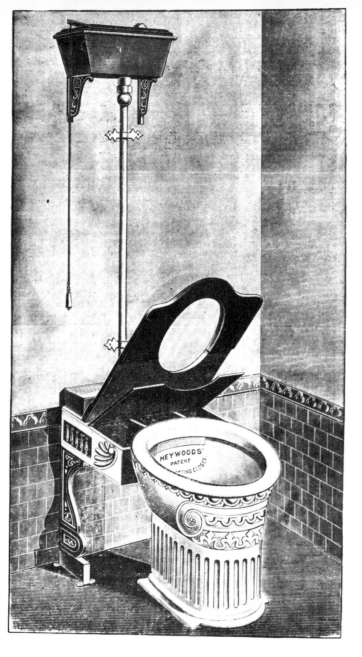

closets ought to be placed in all large hotels and public buildings, and all corporations should encourage their use in domestic dwellings. I cannot speak too highly of these closets; for, if only a proper disinfectant is placed in the chambers, it will assuredly destroy all noxious vapours and prevent the formation of sewer gas, which too often finds its way into dwellings. If this were done, and they became universal in the cities and towns, the corporations would save largely in their sewage departments, for the sewage of a town would arrive at the works deodorized and precipitated." We think that when this invention is better and more largely known a ready sale will be obtained for it, as it appears to be a thoroughly practicable and effective form of a combination closet and disinfector.

Mr. Raimes' Apparatus.

All sewers contain more or less of noxious sewer gases, which are apt to be most foul in the subsidiary soil pipes and drains where the flushing is not good. These noxious gases all tend to rise into the house through the traps of the closets. Mr. H. Raimes, of the Sanitary Engineering Works, West End-lane, Kilburn, has perfected, and is supplying in large quantities, a most useful sanitary adaptation to the water-closet, the joint invention of Mr. H. Barron and Mr. H. Raimes. The essential principle of this invention is that it deals with any noxious vapours at the moment of generation, so that any house fitted with this apparatus, disinfects both its closets and its drains; absolutely preventing the accumulation of any noxious fumes or sediment. The apparatus is in the form of a box, placed at the side of the seat of the closet, which is filled in one of its compartments with a disinfectant, such as Jeyes' Perfect Purifier.

45

The other compartments contain a supply of paper, and a roll of cloth, which may be drawn over the seat, so as to prevent any chill to the person. The disinfectant apparatus may be operated by a push knob, which may be pressed when desired, so as to destroy all odour from the pan, whilst at the same time, a portion is ejected underneath the pan, so that no odour can be perceived when the pan is lifted. Or the disinfectant jets may be operated by the ordinary pull, simultaneously with the lifting of the pan.

The operation of this apparatus is so immediate as entirely to prevent any rising of odour, or possibility of infection to the house.

We understand that Mr. Raimes is deservedly as busy as he can be with the production of his improved apparatus, and that the highest in the land are included amongst his patrons.

An Indicating Door Fastener.

Fig. 1.   Fig. 2.

Some new inventions are so simple and practical that one must feel surprised that they have not been brought out before. This is the case of Ashwell's Patent Indicating Door Fastener. This little device does its work most effectively, and has made such rapid strides in popular favour that within a short time it has been adopted in a large number of hotels and public buildings. This invention will be readily understood from our engravings. By merely bolting the door of the apartment, the word "Engaged" is shown upon the outside of the door, and when the bolt is withdrawn a nickle-plated shield appears in the place of the word "Engaged."

Fig. 1 shows an inside view of the bolt, while Fig. 2 shows an external view of the bolt. The bolt and plate communicate by means of a spindle, which passes through the door, the bolt working in a ratchet wheel, which in its turn actuates the shield or plate. The little device has the advantage of being durable and neat in appearance, and of not being liable to get out of order. It is only necessary to bore a small hole in the door for the spindle to pass through, while it can of course be fitted to doors of any thickness. The fastener is supplied nickel plated or in lacquered brass at comparatively moderate price, and can be warmly recommended as one of those contrivances which contribute greatly to the comforts of life.

Automatic Paper Delivery Machine.

This machine, which has been patented by Mr. Stevens, of Ladywell Road, Lewisham, has been devised to supply toilet paper at railway stations and other places where people mostly congregate. The box is of convenient size, being about 9 in. long, 7 in. high, 4½ in. broad. The box is worked by an ingenious mechanism, after placing a penny in the slot. We hear that this box is finding favour with some of our large railways, as it is well known that these boxes will very often be able to earn a good income where it would not pay to keep an attendant specially.

The untidy, not to say unhealthy, condition of the streets of some of our cities and large towns is the subject of frequent comment, and although sanitary authorities have put forward their best efforts to obviate the complaints made, still there is a wide field for improvement. In London, for instance, thousands of pounds are annually expended in flushing, and in other ways cleansing the wooden, macadam and other pavement.

Horses are employed to a vast extent—the London Omnibus Company alone utilizing no less than 11,000 animals. It stands to reason, therefore, that the time of the scavenger is largely occupied in removing nuisances occasioned by equine traffic.

A gentleman residing in one of the suburbs of London, who has made this matter his special study, claims to have devised means whereby the street orderly will be practically abolished, the pavement cleansed to such an extent that accidents will be largely avoided, the unsightly appearance dispensed with, and the chance of an epidemic of contagious disease being generated very considerably minimized.

The inventor, who has taken out patents in Great Britain, the Continent, America and Canada, contends that if his scheme is brought into general use, the rates of municipalities will be materially decreased; that, whereas streets are, as a rule, washed once a day, it will only be necessary to employ the hose once a week; that the outlay for the "machine" will be a mere bagatelle; that eventually the owners of horses will have every reason to congratulate themselves upon the indirect way they have benefited their steeds; and, above and beyond all, having the satisfaction of knowing that they have lessened the number of serious accidents which are of daily occurrence, owing to the slippery condition of the streets.

The initial cost of introducing the patent will, the inventor further advances, be more than counterbalanced by the returns derived from an effective source which has not hitherto been utilized to the best advantage.

This invention is, we are informed, best described as a "portable horse-bin" and being made of indiarubber canvas or leather, would not cost much to construct or adopt.

A piece of whalebone or steel wire, covered with the indiarubber canvas, is employed at the mouth or entrance of the bag to keep it rigidly open a distance of 6 in. by 3 in., and this framework is bent over to form two hooks to hang to the tail-strap of the present form of harness.

About 4 in. below the mouth and inside the bag is fixed another hoop of whalebone or spring wire, which keeps the upper part of the bag distended, but allowing it below this point to collapse when empty. It is thought that it would be a good plan to hook a little box upon the back of this hoop inside the bag, the lid of which would automatically open as the refuse descended and distribute a disinfectant, in order to prevent any ill effects to drivers. The lid of this box would project outwards in such a manner that it would be caught by the falling dung and automatically opened. The whole concern being buckled to the breeches-strap would, it is thought, in no way inconvenience the horse.

As to finding employment for the bin boys, station, says the inventor, one at each cab stand to empty the horse-bin into a larger bin in the middle of the road, and one at each termini of the omnibus and tramway routes.

The only reasons against the adoption of this system, which he can see, are the initial expense and the fact of the horses' tails being too short to hide the horse-bin to any considerable extent, this latter fault, however, could be, he thinks, easily remedied. Let the horse-bin be made compulsory for public vehicles first, the patentee urges, and private carriages will follow on with them afterwards when the horses' tails are longer.

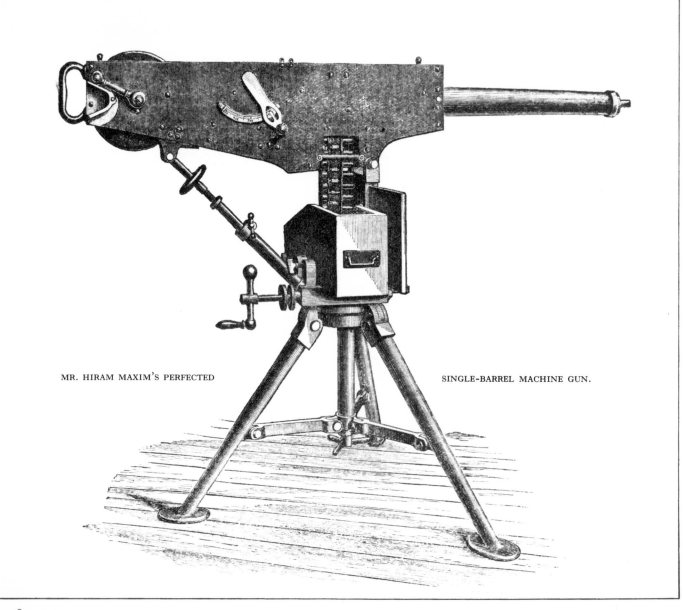

MR. HIRAM MAXIM'S PERFECTED SINGLE-BARREL MACHINE GUN.

CHAPTER VI.

NEW DEVELOPMENTS IN ENGINES OF WAR.

Mr. Hiram Stevens Maxim will be known to our readers as the inventor of the Maxim electric light system, which was introduced into this country at the commencement of 1881, and we may here state that Mr. Maxim's inventions relating to the incandescent lamps were proved in the United States courts to be prior to all others, and that the world owes many important electrical and other inventions to him. Years ago Mr. Maxim conceived the idea of an automatic machine gun, but he did not further develop the conception under the pressure of continuous absorbing work that occupied him in connection with other inventions.

A few years ago, however, Mr. Maxim, arriving in Europe, again took up the idea, which then assumed a definite shape, so that in less than twelve months he had completed in the works established by him at 57D, Hatton Garden, London, an automatic machine gun, which forms an important and thoroughly practical advance in this class of weapon, and a material addition to the means of attack and defence in modern warfare. The Maxim machine gun completely eclipses the machine guns until now in use, from the early Gatling gun and the French mitrailleuse to the Hotchkins and Norderfeld, the fire of which now reaches some 150 to 200 rounds per minute, the operator in most instances having to turn a handle. The Maxim Machine Gun has only a single barrel, but its action is so rapid that it can be made to blaze away at the rate of 600 rounds per minute if desired, while it has the great advantage of automatic action.

This is effected by the recoil, which is utilized as motive power for loading and firing each succeeding round. Thus all the functions of bringing the next cartridge into position, forcing it into the barrel, cocking the hammer, pulling the trigger, extracting the empty shell, and ejecting it from the gun are performed by one recoil.

The Maxim gun is completely under the command of the operator, and by pulling the lever the speed can be so regulated as to fire single shots or volleys of 10, 20, 100, up to 600 per minute. The gun, once started, keeps up a terrific fire without human aid until all the cartridges (with 333 of which the gun can be loaded) are all discharged, and would of course continue to fire, should the gunner in charge of it be killed.

It appears to us that the system described and illustrated here is the most important and the most practical, and its introduction into our army and navy is doubtless imminent. Not only engineers but the most competent military men have expressed themselves highly in favour of Mr. Maxim's invention, and we may state that on Tuesday H.R.H. the Duke of Cambridge and Staff, Sir Fredk. Able and Sir F. Bramwell, inspected the gun and expressed themselves highly pleased with the invention.

We are not exaggerating in saying that this new automatic gun is a triumph of inventive genius and engineering skill, such as is rarely met with even in this age of inventions, and which will secure the inventor one of the first places in the history of the invention of weapons of warfare.

Admiral Popoff's Circular Ships.

The forms and proportions of ironclad ships have undergone great changes in this and other countries since the introduction of armour-plating. Throughout these changes, one principle seems to have been kept pretty steadily in view, that of continually increasing the proportion which the breadth of the ship bears to the length. It is, Admiral Popoff, the distinguished designer of the circular ironclads, avers, merely the extreme form of this principle that he has embodied in his circular vessels, one of which, the *Novgorod*, we illustrate.

Mr. E. J. Reed, C.B., after making a considerable voyage in the Black Sea in one of these vessels in rough weather, reported in a letter to *The Times* that the vessel was singularly steady, even in a rough sea, and that the waves did not roll over the vessel in anything like the quantities which might have been supposed in the case of a ship of so low a side.

The vessels have a very small freeboard, their armoured sides standing only 18 inches out of the water; but the deck, which is 100 feet in diameter, has considerable upward curvature, so that at the middle of the vessel it is 5 feet or 6 feet above the water. Above this deck is a system of super-structures employed as cabins, &c., and in the centre is an armour-plated fixed turret, in which two powerful guns are worked. These guns revolve full-circle with the turret, being worked with much ease by a very simple system of machinery designed for the purpose.

Each of the vessels is propelled by six screw-propellers, placed upon independent shafts, all of which are parallel to each other, and each of which is driven by a separate engine. Mr. Reed asserts that Admiral Popoff deserves the greatest possible credit for the original and inventive thought that the design and arrangement of the vessels exhibit.

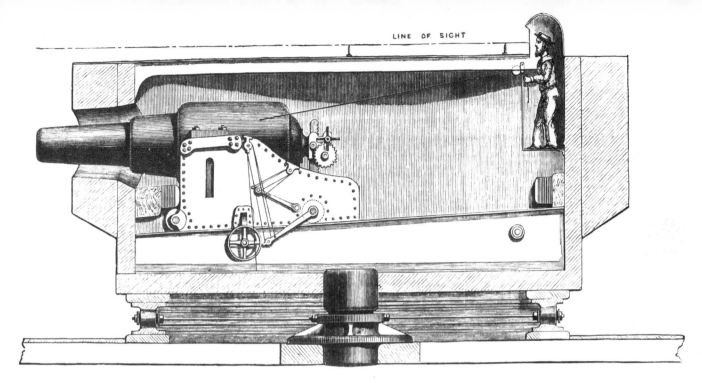

LINE OF SIGHT

The New Gun-Carriages on the *Glatton*.

Our illustration shows a cross-section of one of the two huge new gun-carriages designed by Captain Scott for the iron turret-ship *Glatton*. The captain of the gun is shown in the act of pointing his weapon, which is laid by means of sights placed on top of the turret, and corresponding with the bore of the gun.

The *Glatton's* recent trip to the Nore was arranged for the special purpose of testing her gun-carriages. These stood the severe tests they were subjected to in the most satisfactory manner, and the vessel bore the concussion of both guns firing simultaneously without showing any signs of weakness.

During the trials, witnessed by the controller of the Navy, firing commenced with 55 lb. charges of pebble powder and 490 lb. shells filled with sand. It was continued with charges of 85 lbs. of powder and 600 lb. shot, both guns being fired together,

right aft, loaded with these heavy charges. The success of the trial was so complete that the dummies standing on the decks were not disturbed, and the effect of the blast below resembled only a slight rush of wind.

The *Glatton* is not only armed with her two powerful guns, but with a strong spur protruding under water, for the purpose of ramming an adversary. Her importance can be better estimated when it is borne in mind that the hull of the ship is an impenetrable fortress, with too much flotation to be run over, and too great a strength of armour to be penetrated by any gun as yet mounted in any foreign navy.

Another important feature of this modern ship is the pilot-tower on the hurricane deck, from which the commander, by speaking tubes or by electric telegraph, can direct the movement of the vessel, communicating with the engine-room, the turret, and the steering wheel.

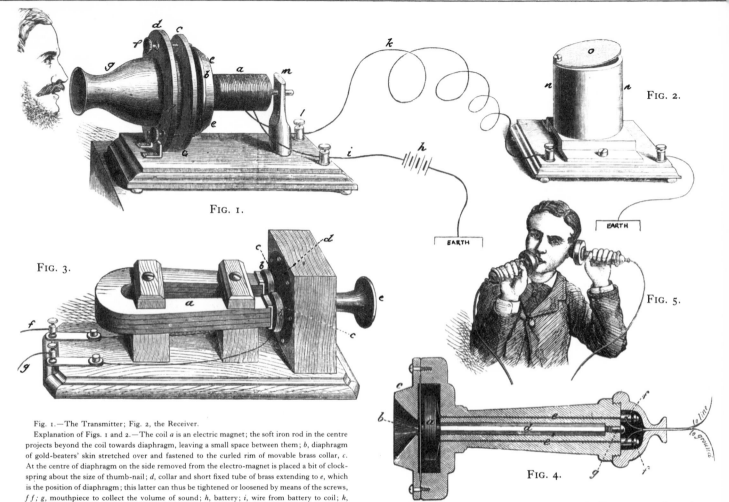

Fig. 1.—The Transmitter; Fig. 2, the Receiver.

Explanation of Figs. 1 and 2.—The coil a is an electric magnet; the soft iron rod in the centre projects beyond the coil towards diaphragm, leaving a small space between them; b, diaphragm of gold-beaters' skin stretched over and fastened to the curled rim of movable brass collar, c. At the centre of diaphragm on the side removed from the electro-magnet is placed a bit of clock-spring about the size of thumb-nail; d, collar and short fixed tube of brass extending to e, which is the position of diaphragm; this latter can thus be tightened or loosened by means of the screws, $f f$; g, mouthpiece to collect the volume of sound; h, battery; i, wire from battery to coil; k, telegraph-wire from coil through binding screw, l; m, pillar holding magnet in place, by means of smaller iron rod, which is fixed to one end of magnet; n, iron tube, inside which is a vertical bar electro-magnet, which attracts and causes to vibrate the thin armature, o.

Fig. 3.—Later form of long-distance Telephone for office use: This is a transmitter as well as a receiver, and here the battery is not required. Its parts are—a, compound magnet; on to each pole of this is clamped a short round piece of bar iron, over which is a bobbin of coil wire, b; c, d, diaphragm of thin sheet soft iron; e, speaking tube; f, telegraph wire; g, line to the earth. The magnet is held in its place by short cross-pieces of wood. The whole is contained in mahogany case to fit in recess of wall, or elsewhere.

Fig. 4.—The portable Telephone (shown in section) which will carry messages five or six miles, and is a transmitter as well as a receiver.

a, bobbin of coil wire round end of magnet; b, diaphragm of soft iron; c, mouthpiece; d, permanent magnet; e, wires, conducting from coil to binding screws; f, the two wires are at the end insulated and bound together in one strand for convenience of use; g, adjusting screw, holding magnet.

Fig. 5.—The Telephone in use. It is preferable to employ two, as represented, to prevent the confusion frequently consequent upon the two persons conversing, when speaking at the same time, which is oftentimes the case where only one Telephone at either end is used.

CHAPTER VII.

THE DEVELOPMENT OF TELEPHONY AND PHOTOGRAPHY.

ADVANCING hand-in-hand, Telephony and Photography usher in a new age. We require no great predictive insight to see the possibilities of this alliance in the future. Already we can visualize using the electric telegraph at the same time as the kinetographic camera, communicating the complete effect of form, colour, motion and sound to points far distant from the scene of action. For the present, in this brief review, we give a conspectus of some of the more significant developments in both fields.

Professor Graham Bell's Instrument.

We give illustrations of the apparatus lately invented by Professor Alexander Graham Bell for the electrical transmission of distinctly articulate sounds to great distances. In general, the "telephone," as it is called, consists of a strong ordinary magnet, to the extremities or poles of which are attached properly insulated telegraphic wires. Just in front of the extremities of the magnet there is a thin plate of iron, and in front of this again, there is the mouth-piece of a speaking tube. By this last, the sounds which it is desired to transmit are collected and concentrated, and, falling on the metal plate, cause it to vibrate. These vibrations, in turn, excite in the two wires electric currents which correspond exactly with the vibrations—that is, with the original sounds. If now the two wires are connected with the ordinary line of telegraph, specially insulated for the purpose, the sounds can be transmitted to any distance, and on arriving at their destination, are reproduced in a precisely similar apparatus. Already there are varieties of the telephone, but this is its essential nature.

In electric telephony the vibrations are not mechanically transmitted; an electric current passes along a wire and reproduces *de novo* a like sound at the receiving end. But in earlier experiments a rod of iron was tried, and subsequently plates; and in some experiments with Mr. T. A. Watson the first articulate sounds were faintly detected. Another modification was tried, and increasing distinctness of articulate sounds resulted; the effects of interposing the resistance of fluids, water and mercury, were also tested.

The form of the receiving apparatus next underwent change. It first consisted of a hollow drum, with the electro-magnet inside, and when the ear was placed on the iron lid the sounds were distinctly heard. Permanent magnets were then introduced, being the primary step to the introduction of portable instruments. Finally the instrument as now manufactured was arrived at, in its handy portable form of a sort of hand-trumpet of some eight inches long by four inches in diameter at its broadest end.

In conversation two instruments are used, one to speak through, the other kept at the ear, as there was found a difficulty in using one instrument only as both the correspondents might be speaking or listening at the same instant; but with two instruments a regular and continuous conversation can be kept up the same as in an ordinary way in a room.

It might be asked how far it is possible for sound to be transmitted. At present no limit has been discovered and it is hoped that the telephone may soon be in use in circuits of all lengths. The longest actual distance through which conversations have been held is stated to be 258 miles—namely between Boston and New York; but in laboratory experiments conversations have been perfectly clear through resistances equal to distances of 6,000 miles. (With respect to the portable telephone, Fig. 5 in our Illustration, although five or six miles carrying distance is claimed for it, yet Professor Bell told our artist that by its aid a message, under favourable circumstances, has accomplished a transit of eighty miles. In fact the powers of the telephone are as yet unknown.)

In the course of a recent evening lecture, at which Professor Graham Bell himself explained his invention to members of the Society of Arts, two telephones were used, one being connected by wire with a neighbouring hotel, and the other having its other terminus in Gough-square, Fleet-street. Questions and answers were sent through these apparatus (one of the experimenters being the Earl of Caithness) and the responses were distinct, and almost immediate.

A New Type of Telephone.

The telephone is coming rapidly to the fore, and yet there is still room for improvement in this particular field. As is well known, the present telephone monopoly is, in a great measure, due to the "Edison" and "Bell" patents. The Equitable Telephone Association Ltd. has, however, got into working order, and we now illustrate the invention of Mr. A. A. Campbell-Swinton, which is not covered by the above patents.

The instrument comprises a direct-acting multiple microphone transmitter, self-contained call-bell, push-button and automatic switch, mounted on a polished teak wood base-board, and two electro-magnetic receivers.

The transmitter consists of a lead frame adjustably suspended by india-rubber pieces, so as to be unaffected by external vibrations or tremors. On a horizontal platinum wire stretched across the upper part of the frame are strung a number of vertical pendulous carbon pencils, which rest lightly at their lower extremities against an insulated horizontal carbon block fixed across the back of the frame, the whole forming a very powerful multiple microphone. (See the central rectangle in the accompanying engraving.)

The microphone sensors are so sensitive that they are readily actuated by the direct impact of the atmospheric sound waves on themselves alone, without the intervention of any diaphragm,

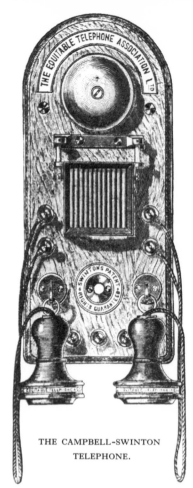

THE CAMPBELL-SWINTON
TELEPHONE.

tympanum, or auxiliary sound-receiving surface of any description, or the aid of any mouthpiece or voice tube. This direct action has the advantage of rendering the articulation of the telephone remarkably pure and distinct.

A formal guarantee against any claim for any alleged patent infringement is given with each set of instruments, and the company will indemnify all purchasers and users against any such claim.

Binko's Portable Telephone — Tricycle Communication for the Army in the Field.

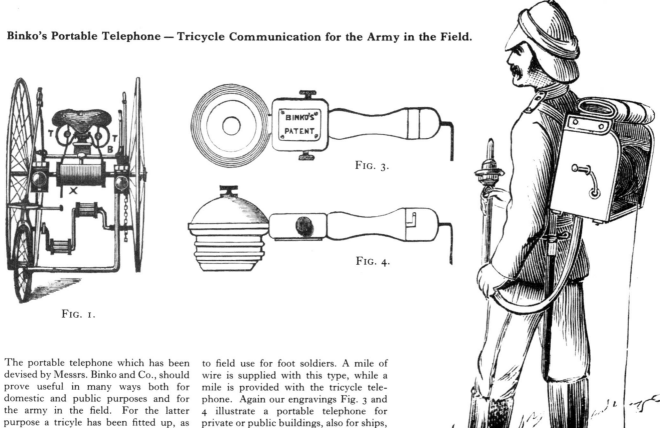

Fig. 1.

Fig. 3.

Fig. 4.

The portable telephone which has been devised by Messrs. Binko and Co., should prove useful in many ways both for domestic and public purposes and for the army in the field. For the latter purpose a tricyle has been fitted up, as shown in our engraving, Fig. 1. T T are a pair of telephones suspended under the seat. B is the battery box, and X a coil of wire wound round a bobbin on the axle. The wire is paid out and rewound as the machine goes along. It will be seen that the whole apparatus is very compact and portable.

Fig. 2 shows a modification applied to field use for foot soldiers. A mile of wire is supplied with this type, while a mile is provided with the tricycle telephone. Again our engravings Fig. 3 and 4 illustrate a portable telephone for private or public buildings, also for ships, railways, &c. In this case, no fitting whatever is required, and the telephone can be moved from one place to another without trouble, like a portable electric bell. Our drawings are one-third actual size. Receiver, transmitter, bell, and push are combined in small compass. It is claimed that they can be joined to an existing electric bell system if desired.

Fig. 2.

Photography with Sensitive Material on Rollers — a Recent Innovation in the Dry Process.

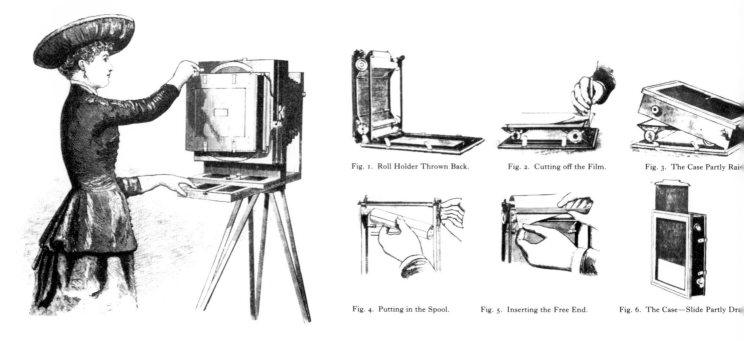

Fig. 1. Roll Holder Thrown Back.   Fig. 2. Cutting off the Film.   Fig. 3. The Case Partly Rai[sed]

Fig. 4. Putting in the Spool.   Fig. 5. Inserting the Free End.   Fig. 6. The Case—Slide Partly Dra[wn]

The enterprise of the Eastman Company in introducing so noteworthy an invention as their roll holder, and the excellent sensitive paper film used with it, is illustrative of the characteristic of push and energy so often displayed by American inventors; we bespeak for their improvement an important future, and consider it an advance in the art of photography which will be welcomed both by amateur and professional alike.

The universal success and appeal of photography has been accomplished by the introduction of the dry plate but there have been many who have been prevented from practising the art by the weight of the apparatus and material, which has to be carried about even to make a few pictures. The weight of the glass is such a serious burden, especially in the larger sizes, that even the most enthusiastic may be discouraged after a few trials.

By reason of several recent improvements it has been found possible to prepare paper of fine and close texture, with an even coating of the photographic emulsion, so perfectly that positive prints made from the paper negatives show no grain in the half-tones, and are as clear and perfect as if made from glass. The paper is wound upon a wood spool, arranged for use in an instrument termed a roll holder, the principle of which is to draw the sensitive paper from the supply spool at one end over an exposing platform.

A guide roll is placed at each end of the platform, the one on the right (Fig. 2) being termed a measuring guide roll, in which is a longitudinal slot used as a guide for the point of a knife in cutting off the exposed from the unexposed paper. The roll has a pin at one end for operating a flat spring, making an audible click. As the circumference of the roll is one-fourth of the length of the picture, four clicks are sounded, and a counting indicator makes one revolution when one exposure has been wound up.

The exposed sheets, as they are cut off, can be developed several at a time, in one tray. The developer is sold already mixed, thereby insuring to the novice success at the outset.

After the negative is fixed and dried, positive prints may be made from it in the usual way, but to quicken the process, oiling the paper with castor oil and a hot iron which renders it transparent.

After exposure a new spool of paper may readily be inserted to take the place of the used spool. Paper to the correct width and length is sold by the Eastman Company in light-tight cardboard cartridges, specially for the purpose.

It need not be stressed that the primary advantage of paper over glass is its extreme lightness. An 8 × 10 apparatus, complete, with camera, lens, roll holder for 24 exposures, tripod and case, weighs 28 pounds less than a glass equipped outfit. Such a saving makes the taking of large photographs attractive, and enables the amateur to obtain panoramic or other views of inaccessible regions with considerable comfort.

The compact way in which the negatives can be stored should not be overlooked: they can be kept in books, thereby affording as easy a means of reference as if they were in a photographic album.

Not only is the softness and delicacy of the paper negatives especially noticeable; there is a particular brilliance in the high lights too. There is also an added advantage in that the retouching of paper negatives is more easily carried out on paper than on glass, for the back of the negative is worked upon by a pencil; any mistake can be easily erased.

There is little doubt that the improvements embodied in this process will go a long way to bringing the photographic art within the compass of a wide circle of enthusiasts, especially the fair sex, who, as our illustration conveys, experience no difficulty either in carrying or in operating the roll holder. The rapid strides made by photography in the last few years, with the complicated paraphernalia of wet-plate processing already a thing of the past, allow us to look forward to the continued diminution of the bulk of the apparatus. It seems not unlikely that the truly "handy" camera is actually within sight.

A New 5lb Portable Camera.

This camera loaded with sheaths, weighs only a trifle over 5 lb. It is as simple as it is ingenious.

It has a rising and cross front which enables the operator to dispense with the usual excess of foreground, and another advantage claimed for it is that the full size picture can be seen both horizontally and vertically, and is perfectly distinct in the brightest sunshine. These last are two important features which, we believe, have never before been attained. The plates are changed easily and rapidly by pulling out a knob on top and then returning it to its original position. The same movement also registers the number of plates exposed. By this effective automatic method of changing the operator is able to take several pictures in quick succession, the register obviating any danger of confusion as to the number of unexposed plates remaining. The shutter, which is specially designed for this camera, is somewhat unique. It is fitted with Hill and Adams' patent regulator, and has no vent-hole to admit dust. Another advantage is that the whole shutter being made entirely of suitable metal, it is reliable and effective for hot and tropical climates. It permits of exposures varying from 1 to 1-100 second, and any speed between. It is made on the "ever set" principle; it works in the diaphragm slot, and has an automatic self-capping device, and, which is another important feature, has a direct movement across the lens. This, and the easy manner in which the shutter is released, avoids any chance of vibration. The focussing may be done up to the time of exposure. There are two finders, one showing the horizontal and the other the vertical picture, both full size, and the colours of the pictures shown are unusually vivid. Perhaps the most radical improvement is that the reflecting mirror and ground glass are entirely dispensed with. The illustration shows how the focussing is accomplished. The shutter release is so conveniently arranged that by merely turning the milled head screw it can be immediately let off, thus ensuring the exposure being made at the moment the picture is seen to be in focus.

The Photographic Necktie.

Where will the progress of instantaneous photography end? In view of the admirable results obtained by scientists, inventors have for several years been setting their wits to work to devise small apparatus for allowing amateurs to take photographs without anyone seeing them do it. We have already heard of the photographic opera glasses and hat; but here we have something cleverer, and designed to meet with great success among practicians; it is a necktie provided with a pin. The latter is a lens, and the necktie is a camera. When any one approaches you and speaks to you at a distance of two or even three feet, you press a rubber bulb concealed in your pocket, and you have the portrait of your interlocutor.

This ingenious little apparatus, with which also general views may be taken, was devised by Mr. Edmond Bloch.

Fig. 1 represents the photographic necktie, and Fig. 2 gives a front view of it as it is to be worn by the operator, the metallic camera, which is flat and very light, being hidden under the waistcoat. Fig. 1 gives a back view, the cover of the camera being removed to show the interior mechanism, comprising six small frames which are capable of passing in succession before the lens, and which permit of obtaining six negatives. The apparatus is operated as follows: The necktie having been adjusted the shutter is set by a pull upon the button, which

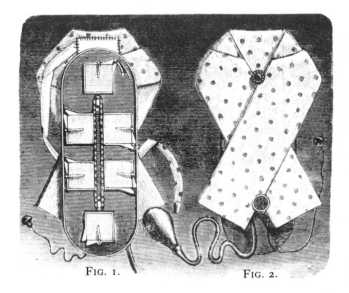

FIG. 1.   FIG. 2.

passes under the waistcoat. In order to change the plate, it is necessary to turn from left to right the button which has been introduced into a button hole of the waistcoat, and which simulates a button of that garment. This button must be turned until the effect of a locking is perceived. This puts the plate exactly before the lens. In order to open the latter, it is necessary to press the rubber bulb, which has been put into the trousers pocket. The rubber tube serves to transmit the action of the hand.

The photographs the camera produces are about $1\frac{1}{2}$ in. square, and are sufficiently sharp to allow the portraits to be recognized. We predict a great demand for this invention.

CHAPTER VIII.

IMPROVEMENTS IN THE AGRICULTURAL FIELD.

No less than elsewhere, in agriculture the ingenuity of the inventor is brought daily into play. The introduction of the Steam Traction Engine, to name but one development, has transformed the rural scene, bringing much-needed help to the farm-labourer in a multitude of different applications. If the introduction of mechanical aids is perhaps slightly less forward in the countryside than in the towns it is only because the knowledge of the existence of such machinery is less readily acquired on the farm than in the Metropolis. With the increase in the dissemination of knowledge, this disparity may shortly be expected to be overcome. We give here a few examples of new aids to the farming community.

We avail ourselves of the opportunity to place before our readers an illustration of an unique and ingenious contrivance. This is Mr. Jens Neilsen's Automatic Milking Machine, exhibited at the recent Dairy Show in London.

The Machine is of particular importance to dairy farmers who have difficulty in procuring skilled milkers. Obviously, any lad can manipulate the contrivance illustrated.

The action of this mechanical milker has been devised to closely imitate the sucking functions of a calf's mouth, without causing irritation to the cow. In this milking machine the desired action has been obtained in a perfect form, all four teats being milked simultaneously by two pairs of elastic and feathering roller segments, having suitable rocking, approaching and receding movements. The teats are squeezed from the upper ends down to the bottom. When one pair of the rocking segments approach each other, squeezing the two teats on the right side of the udder, the other pair of segments on the left side, recede from each other and conversely. The operator turns the handle shown, situated an arm's length from the right side of the cow, and connected with the main shaft by a link chain. The machine rests in a self-adjusting frame, suspended from the cow, so as not to be affected by any movements the cow may make during the milking.

The machine is put in place in a few seconds, and removed simply by a turn of the hand. The milk flows through a funnel into the milk can, and the operator is thus able to see when the

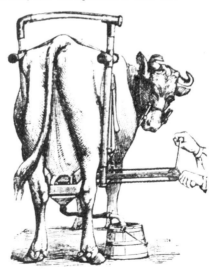

59

cow is milked clean, that is to say, when no more milk flows. It is claimed, from observation and experience, that cows like the process and keep perfectly quiet during the mechanical operation, and even in the case of cows with sore teats, it is stated that they keep quieter than when milked by hand, for the reason that the mechanical pressure is more even and softer, and throws less strain on the milking organs. Unquestionably the employment of the machine makes milking a cleaner and easier work.

A Mechanical Sack Turner.

The construction and action of this device are so simple that they want little description beyond what is conveyed by the accompanying drawings. The machine consists of two cross bars of steel, or iron and brass, hinged to a stand plate which can be screwed to the floor. The bars can be adjusted to suit

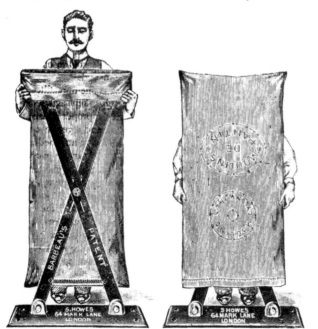

FIG. 1.   FIG. 2.

various widths of sacks by means of the slide and screw arrangement shown in the centre. The operation of turning the sack is shown in Fig. 1. The upper end of each bar terminates with four small rollers, and some three inches at the mouth of the sack having been turned back, the lap of same is placed over the rollers, as shown, and drawn downwards until the sack is completely turned, as represented in Fig. 2.

Machine for Broadcasting Fertilizers.

We illustrate an ingenious new machine for broadcasting nitrate of soda, basic phosphate, lime, soot, cake dust, sulphate of ammonia, and chemical fertilizers; for sowing broadcast wheat, oats, barley, peas, tares, sanfoin, clover, and other seeds, and distributing liquids of every kind for destroying insect and fungoid pests on turnip and other ground crops.

The machine can be adapted for sanitary purposes, distributing disinfectants in streets, and for sanding and salting roads.

CHAPTER IX.

LOOKING INTO THE FUTURE.

PASSING in review the history of development of electricity, are we not convinced that about 50 years ago no one would have believed a prophet of the future foretelling all the present wonders of the spark? And what of all the other wonders, newly vouchsafed to us by the scientist and inventor?

We do not burn the wizard any more but we must ruefully concede that often they are still treated as fools. We must learn to stop this. We must no longer summarily denounce scientific predictions, even of the most revolutionary character. All we are allowed to say is this: "Who can tell? This may be very well possible." Thus we do not risk to engage in a dispute which sooner or later will be decided against us by the facts. In these last pages of our review of our century of progress we give a few examples of what the future may hold.

The subject of a cross-channel railway, or other Anglo-Gallic communication route, has been under serious discussion since 1842, at least. Mr. E. Pearse, of Devon, published a letter on the subject in the *Railway Times* in that year in which he suggested the construction of an iron tunnel at a vastly less comparative cost than the Thames tunnel, which was of brick. In 1848 M. Ferdinand, engineer, submitted to the French Academy of Sciences a proposal to construct a floating tunnel from Calais to Dover, to carry the wires of the electric telegraph, and large enough to be transversed by small passenger locomotives.

Earlier, in 1838, Mr. Rettie, C.E., of Glasgow claimed to have suggested submarine railways similar in construction to the plan of Mr. De la Haye of Liverpool, whose project in 1845 for submarine railways between Liverpool and Birkenhead, North and South Shields, and Dover and Calais, was published. Mr. De la Haye's plan simply called for the construction of wrought iron tunnels to be sunk on the sea-bed in sections.

"It will be admitted that to construct such a tunnel would be an easier matter than to build iron vessels, as it would be the same shape the whole length; then to sink it on the bed of the water would be the work of only a few hours for each division of 400 feet in length. Perhaps the part of the work which will appear the most complicated will be to connect the divisions under water. The operation will be attended with no extraordinary difficulty to those who can remain during half an hour in deep water. As regards that part of the tunnel which would be near the shore, it would be sunk under ground and covered with stones fastened together, so as to render them immovable. Then the railway will be formed in forming the tube; there would be no hills to cut through, valleys to fill up, or arches to build. In short, the sum total of the work is comprised in the tunnel itself."

It is evident how widely the desire for the fulfilment of this great international work has extended, and what a hold the idea of the practicability of a submarine railway from Dover to Calais has got on the minds of practical men. Indeed there can hardly be a doubt that it will yet be *un fait accompli*; and therefore it seems a pity that its benefits should be deferred beyond our own time. The invisible highway (if that term be admissible) would link together the two greatest nations of the world, already so happily allied. In these days of angle-iron and boiler-plate, and when such a thing as the girdling of the earth in 40 minutes, as Puck talked of, would be readily undertaken by our telegraphers, this great project should not lightly be discarded. It becomes rather a duty first to see whether the thing proposed is worth having; and, if so, whether it can be brought within reach, by divesting it of the difficulties in which it may seem to be enveloped.

M. HOREAU'S SCHEME FOR A CAST-IRON TUBULAR TUNNEL, LYING ON THE SEA-BED. THE COST OF THIS PROJE

The Submarine Railway of M. Horeau.

The communication between England and France, which is daily becoming a more and more important object, has received a most valuable acquisition in the laying of the Submarine Electric Telegraph. Thus far the transmission of intelligence from one country to the other is instantaneous and complete. To consummate the international union, however, means of the conveyance between the opposite shores, so as to avoid the perils and uncertainties of passage by sea, remain to be provided; for, with all our scientific appliances, we have yet occasionally to read in our morning journals such a paragraph as—"At the time of our going to press the French mail had not arrived at Dover."

The communication has long been a favourite scheme with projectors. Bridges have been proposed by some; by others it has been suggested to tunnel the earth beneath the sea, which the works now executed for mining purposes seemed to render possible; but the difficulties would be insurmountable, and the cost enormous. Another projector, M. Hector Horeau has, however, just appeared in the field, with what he allows to be a bold plan, but which appears to him to hold out the requisite guarantee for so important an undertaking.

M. Horeau's project consists in crossing the English Channel, 21 miles in extent, by means of a tube, or tubular tunnel, made of strong plate iron, or cast iron, lined and prepared for that purpose; and which, placed at the bottom of the sea, should, besides the path for the surveyors, contain the two lines for the trains which would run within this tube.

The slope given to the submarine railway, M. Horeau considers, would admit of a motion sufficiently powerful to enable the carriages to cross the Channel without steam-engine. The greatest depth of the sea in the middle of the Channel will admit of the construction of inclined planes, by means of which the train would be enabled to reach a point where a stationary engine or atmospheric pressure might be employed in propelling the train would be enabled to reach a point where a stationary

These tunnels beneath the sea would not prevent navigation: two lighthouses might be erected at the entrance of the tubes; also several smaller ones between the lighthouses of France

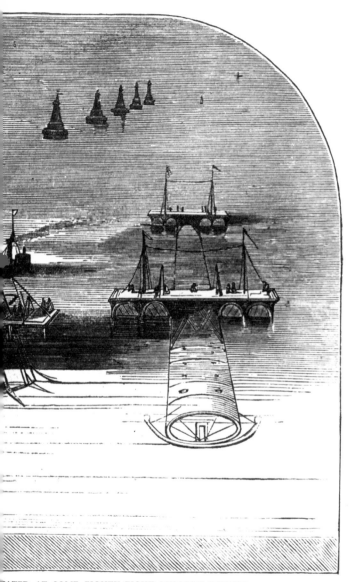

ATED AT SOME EIGHTY-EIGHT MILLION POUNDS.

and England. These beacons, which may bear the names of the different nations of the earth, should be lighted up at night, and would indicate outwardly the position of the submarine railway, so that mariners should not cast anchor near it, as the tube might be damaged.

The day and night lights of the lighthouses should be transmitted through the tube (covered internally with a coating of enamel or lead) by means of reflecting metal plates. The upper part of the tube should have some strong glass windows placed at equal distances, and gas, which would complete the lighting between the beacons: the carriages might also be open, or have glazed roofs, to enable the passengers to profit by the various lights. According to an estimate made, the cost might amount to about £87,400,000.

Mr. James Wylson's Project.

Mr. James Wylson proposes a cross-channel railway of a different kind. As the annexed engraving shows, he proposes to situate the tunnel at a uniform depth from the surface by means of ties below (and buoys above, if necessary) at suitable intervals. The continuation of the tunnel into the shore at either coast would be dispensed with; in order that it would have a partial freedom of motion (*floating*, as it were, in the water of the channel) it should terminate with solid ends before reaching the shores. From the end portion at either termination of the tube a shaft or staircase would rise, terminating above the surface of the water in a railed and buoyant platform, with roof and other appliances for shelter and comfort.

The trains might alternate from opposite shores, and the engine retaining its position at one end of the train, pushing one way and pulling the other, —there being only one pair of rails. Thus, the subaqueous arrangement would be very simple. Electricity should, if at all practicable, be the motive power; and it is pretty evident that it might be the lighting medium. The top and sides of the carriages should be of glass, and the light in the tunnel continuous, that those who rode might read.

Supported from the buoyant base of each ventilation shaft might be one or more bells, so arranged as to be swung by the action of the wind and sea, and thus give notice to mariners

MR. WYLSON'S PROJECT FOR A "FLOATING TUNNEL".

by night of their proximity to the vault beneath them; or, were it considered desirable, the upper parts of the shafts might be illuminated so as to indicate the position of the tunnel at night.

It is not to be assumed that the sea immediately below the surface is in a perfectly quiescent or stagnant state; it has been ascertained that the oscillating motion caused by the agitation on the surface may extend sensibly to full 100 feet in depth. For this reason a partial freedom of motion is provided for. That mobility would not be comparable with safety to the trains were the ordinary rails only to be employed; it is therefore proposed that there should be a guard rail set up on each side in a groove in which the extended ends of the axles should run. It is not to be supposed that the degree of motion spoken of would be anything like, in extent, to that experienced on board ship.

A Novel Implement of Destruction.

Mr. James Nasmyth has offered to the nation a device "for destroying by one masterly blow, the largest ship of an invading enemy".

The principles on which the arrangement and construction of the Floating Mortar is based, consist in the first place of a monster self-exploding shell, so arranged as to explode on having its breach end crushed against the breach of the Mortar; the self-exploding cap being situated there, as will be seen on reference to the illustration.

In order to enhance the destructive effect upon the enemy's ship, the shell is so far submerged as to tear its way into the enemy six feet under water line. No ship that has ever been built, whether of wood or iron, could survive the fearful hole which a Monster Shell, exploded under such circumstances, would produce.

This class of vessel is chiefly designed for defence against invasion, and would not have to act against an enemy, probably, at greater distances than one or two miles from our shore. It could speedily return for another shell, the means for lodging which in the chamber of the submerged Mortar are most simple, but not needful at present to describe. It is conceived, however, that the total destruction of one enemy ship at each trip would be sufficient service.

The draught of the engine-furnace would cause perfect ventilation for the crew, which need not consist of more than three or four handy men.

MR. JAMES NASMYTH'S "FLOATING MORTAR", WITH MONSTER SELF-EXPLODING SHELL.

An Impregnable Fortress.

Mr. William John Hall's Impregnable Iron Fortress has recently been drawn to the attention of the public. That public-spirited gentleman has pointed out the dangers to which the community would be subjected in the event of the sudden attack of an enemy. He points out that there can be no doubt that the certainty of our being enabled to resist an attempt against London by fluvial invasion is the very best security against any hostile attack being attempted.

His fortress is illustrated here so that readers might form an independent judgement of Mr. Hall's claims.

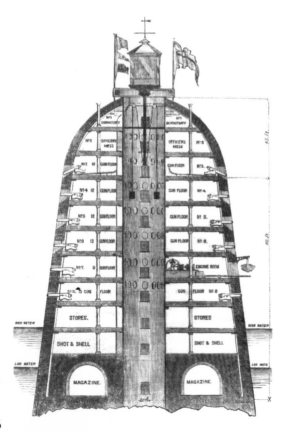

The nature of the proposed fortification may be readily understood from the accompanying engravings. The diameter of the structure at high water mark will be 120 feet, and its height from that line about 130 feet—the total elevation being 240 feet. There are portholes for seventy cannon, twenty-one of which can always be brought to bear upon a given point. The internal space will accommodate a garrison of fifteen hundred, whilst, upon the assumption that one third of that number will suffice for all ordinary purposes, Mr. Hall suggests the propriety and social policy of training a thousand destitute lads from our streets into useful members of society, as sailors, engineers, or mechanics. The edifice will have for its base an enormous caisson, sunk through the sand to the solid substratum, and filled with concrete, whilst the exterior wall will be composed of cast-iron blocks, each of five tons weight, so dovetailed and amalgamated by fused metal as to form a homogeneous wall of iron, two feet in thickness, and absolutely impenetrable against all known projectiles.

The weight of the fortress will be, in iron, about 32,000 tons, making, with wood and concrete, a total weight of about 110,000 tons. Its magazines are to be formed in the solid concrete, and placed below low water mark. It will be surmounted by a lighthouse, capable of being lowered into the capacious airshaft, 20 feet in diameter from the base to the summit, and which will serve, among other important purposes, for pumping water by steam power from an artesian well, and raising powder, shot, and shell from the magazines, and stores of all kinds, from the lower to the upper floors. The building will be heated by steam passing through the hollow columns supporting the floors.

But what strikes us as the great peculiarity of the scheme is the utter absence of all sinister or personal objects on the part of its promoter; for Mr. Hall puts forth the result of his patriotic ingenuity and perseverance entirely at his own cost; seek no patent; pretends to no infallibility; courts suggestions of improvement; desires only that Government shall do its duty efficiently and economically to the vast interest of British commerce; and, failing in that effort (if the possibility of failure in such an effect can be anticipated), offers to contribute largely towards the patriotic fund for accomplishing the great object of this laudable ambition.

IMPREGNABLE IRON FORTRESS FOR THE COAST AND RIVER DEFENCE OF THE UNITED KINGDOM.

A New Aerial Machine.

Flying machines are by no means a new idea, and the failures of previous inventions in this direction are to be taken as establishing the impossibility of aerial flight, even partially, by man. Whether the machine here described and illustrated, solves the problem, we would very seriously question; but the fact that the *Scientific American* has given it a place in its columns may be taken as

DR. W. O. AYRES' INVENTION.

attributing some value to the invention. This new plan for aerial navigation has been designed by Dr. W. O. Ayres, of New Haven, Connecticut.

The motive power, it is proposed, is to be compressed air, which is intended to be condensed within the two drums, seen in the engraving; the air also fills the hollows in the tubular framing of the machine.

The air will be condensed under a pressure of, say, some three thousand pounds to the square inch. The drums and tubes are expected to hold air enough to drive the engines and attached propellers for several hours. The author gives the following additional particulars. "The plan and form which we suggest is not designed or expected to be by any means exclusive. The illustration shows it very clearly, and we believe that a machine constructed as here represented can do its work successfully. The propellers may be made to present a much greater extent of surface than the artist has drawn; the only thing for which we contend is that the principle shall be maintained. In order to afford support for our two systems of propellers, we must necessarily have vertical posts and horizontal bearings as well: that is, a table-like frame.

A construction measuring 4 ft. by 3 ft., supported by four legs 4 ft. in height, will give us the required space, and if made of steel quarter-inch tubing, will have all the strength needed. The rider sits in a seat like that of a bicycle, suspended by steel wires from the top frame, with which his shoulders are roughly about on a level.

The four horizontal propellers have their bearing on the vertical posts just below the upper frame, thus bringing the lifting power as far above the centre of gravity as possible. The vertically moving propeller revolves on a shaft behind the shoulders of the rider, midway between the side bars of the top frame. The air cylinders are two, for better division of weight, but are so connected that practically their air mass is one. A driving engine is attached to each cylinder, but the two work synchronously, and the master regulating valve is controlled by the rider's left hand.

The engines are so geared as to propel the upper horizontal fans, which have been already mentioned. The rider's feet rest on pedals like those of a bicycle, and by suitable connection actuate two horizontal fans as shown, so that the entire strength of his lower extremities can be brought to the assistance of the compressed air in the work of lifting.

There remains only the vertically revolving propeller; this is easily driven by the right arm of the rider, and the gearing, as shown in the engraving, is very simple. An ordinary crank handle is conveniently placed for his grasp, and he drives the fan by direct motion . . ."

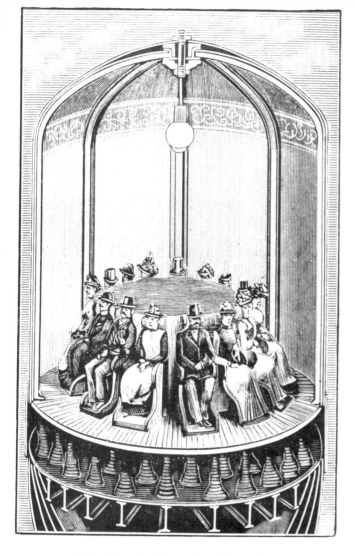

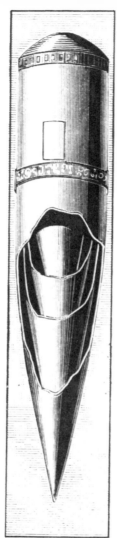

A Machine for Sensational Emotions

Monsieur Carron, a Grenoble engineer, has invented a machine which will be the delight of the lovers of sensational emotions.

The inventor has thought of those who are fond of such strong sensations as are experienced upon a swing, or upon mountain slides, particularly where the descent is rapid. To increase this emotional feeling, he proposes to give to the public the impression of a vertical fall of 1,000 ft. (300 metres, the height of the Eiffel Tower). The project is a new one, and practical, if M. Carron's calculations are correct. At the end of a fall of 300 metres, the velocity acquired is 77 metres per second. Very rapid trains only travel about 30 metres per second, so never has the human species travelled at this rate before. The sensation that will be felt is that of giddiness; it is easy to fall the distance named, but difficult to pick one's self up again after such a fall. Herein lies the inventor's secret. Construction: A cage in the shape of a ball or mortar-shell, in the head of which is a chamber 3 metres diameter by 4 metres of height, capable of holding 15 persons comfortably seated in well-stuffed arm-chairs, circularly arranged; the floor is a mattress with springs of 50 centimetres in height; the issuing or narrow point of the bomb beneath is fitted with a series of cones interlocking each other. The total height of the apparatus is 10 metres, and weighs 4 tons. From the top of the Eiffel Tower this gigantic shell is allowed to fall with its load, but is not crushed by coming in contact with the earth, for the inventor has provided a large basin or pond full of water—or, more properly, a well, excavated in the shape of a champagne glass, with a diameter at top of 50 metres and a depth of 55 metres; at the depth of 28 metres to the bottom the diameter is 5 metres. The shell, on its arrival, is received by this yielding cushion, and displaces 30 tons of water; the wave produced by the fall is spent at the edge of the basin. According to M. Carron, the reaction of the shock to the passengers is annihilated. The shell after its fall would float, and a landing stage would be put to it for the passengers to disembark, then the shell is again raised by an apparatus to the top of the tower to begin again the emotional descent. The price is fixed at 20 francs per passenger, a moderate sum for such sensational enjoyment.

The Flying Man's Parachute.

Mr. de Groof, the Belgian, who was killed in the unfortunate accident after a balloon ascent from Cremorne Gardens had been employed several years in constructing for himself the apparatus with which he believed it possible to imitate the

MR. DE GROOF IN THE AIR.

flight of a bird. The general plan of the apparatus was an imitation of a bat's wings, the framework being made of cane, and the intervening membrane of stout waterproof silk. The wings were altogether 37 ft. long, with an average breadth of 4 ft. The tail was 18 ft. by 3 ft. These wings were inserted into two hinged frames, attached to a wooden stand, upon which the aeronaut took his place. He had three levers, which he worked by hand, to give his machine propulsion or guidance, as might be required. His theory was that, having started from a given height, he could manage his descent so as to reach the earth by a sort of inclined swooping motion, without risk of concussion.

About a year ago M. de Groof made an attempt to descend

from a great height on the Grand Place at Brussels. The effort was a failure, but he was not hurt. He came this summer to London, and one evening went up by the balloon from Cremorne with Mr. Simmons, having his machine attached to it. On that occasion he descended safely from a height of 300 ft. or 400 ft. in Epping Forest. A second attempt proved abortive, the machine not working properly, so that Mr. Simmons refused to take him up.

On the fatal evening of his last attempt it was intended to let the parachute descend in the Thames. M. de Groof was able to detach himself from the balloon when he pleased. He had arranged with Mr. Simmons to let the balloon be within a certain distance of the ground for this purpose. The balloon was accordingly lowered from 4,000 ft. to 300 ft. above the ground.

They were drifting near St. Luke's church, not much above the height of the church tower. De Groof seems to have detached his machine from the balloon immediately afterwards. The apparatus instead of inflating with the pressure of the air, collapsed, and, turning round and round in its descent, fell with de Groof in it.

The balloon rose and went on, crossing London in a north-easterly direction. Mr. Simmons swooned in the car, and did not recover his senses till he was over Victoria Park. He travelled into Essex, and came down with his balloon on the railway, just in front of a train, which the engine-driver stopped in time to prevent another accident.

It is melancholy events such as this that underline the fact that the path of the inventor is not without its hazards.

Aerial Propulsion by Explosion or Impulse. A Novel and Remarkable Theory.

Edwin Pynchon, M.D., of Chicago, has been considering the problem of aerial flight and in this year of Grace 1894, appears in print to expound his theory of how the whole problem may be solved by the use of nitro-glycerine or other high explosive agent.

The Doctor firmly believes in the feasibility of the aeroplane for mechanical flight and agrees with most of the conclusions demonstrated by Professor Langley and Mr. Hiram Maxim, but he argues the faster the craft is propelled through the air the greater will be its lifting capacity. Steam or electricity do not appear to embrace sufficient possibilities for the Chicago inventor, hence his advocation for the use of an explosive as a propulsive agent.

Without a sufficient velocity the supporting power of the air cannot be utilized, he says. So long as the momentum of the vessel holds a proper relation to its specific gravity, so long, urges the Doctor, will horizontal motion be made in safety.

The inventor declares his belief in the employment of intermittent explosions or impulses for propelling the machine in preference to a lesser uniformly maintained power of propulsion. Subsequently he proclaims nitro-glycerine to be the ideal "portable force," although it is allowed that dynamite is safer to handle: this leads him up to a discussion of their composition and comparative properties.

The most important part of the whole device, he tells the reader, is the mechanism for using high explosives. From the magazine room, which is well forward, there extend rearwards two solid oval or grooved pipes of about one inch calibre, each terminating by passing through the upper edge of a concave detonating plate, preferably made of some copper alloy, which plates are placed one at either side of the stern of the vessel exterior thereto and near the horizontal centre of the ship's weight. Cartridges are to be automatically fed to these pipes at suitable intervals by a mechanism similar to that found in a magazine gun.

The cartridges when delivered into the pipes are to be shot by pneumatic pressure to the outer opening at the rear of the vessel and there exploded in the concavity of the detonating plates by aid of the electric current.

He tells the public that he contemplates building a structure of about double the size and weight of the Maxim machine. In order to give lateral stability, the planes, he remarks, had better extend from the vessel's centre slightly upward toward their outer edges and proper rudders must be providing for steering. The propellers are to be placed one at either side of the vessel near its forward end and beneath the aeroplanes. It is to be understood that the use of the propellers will only be required when going at low speed and during which time the rudders will be of but little use, but as the speed increases the rudders will be of more and more avail, until, it is contended, their operation will, in fact, be the only means required to attain and maintain the elevation and direction desired. If at any time necessary to turn about in face of a very strong wind, the vessel can, we learn, be powerfully veered by using for a sufficient time only one side of the explosive propelling service.

In using such air ship, it is explained, after some degree of ascent and forward motion has been made by use of the propellers, a pair of cartridges of low power are caused to detonate. These cartridges are to be fed to their respective tubes and by a moderate force of compressed air slowly shot through the same. When both cartridges are in proper position for firing a weak electric current is closed which thereby automatically throws a switch and allows passage through the cartridges of a sufficiently strong current of electricity to fire the same. At the start when the speed of the ship is moderate light charges should be used and the said charges increased gradually in size or strength until the maximum speed of the vessel is attained, which may then be maintained by using the maximum charge with such frequency as practice teaches to be best and which would be more frequent when adverse winds are being encountered than when going with a favourable breeze.

In aerial travel the great desideratum is, we are told, ceaseless and rapid onward motion, and at an altitude of from 500 to 2,000 ft. the best results should be attained. It is quite probable, says the Doctor, that a speed of 150 or 200 miles an hour can be easily had and will, in fact, be necessary in order to ensure a commercial success. He has estimated, with a ship of the size mentioned, that afer a full speed of 200 miles an hour has been attained it can be maintained by the explosion every five seconds of a pair of 60 per cent. nitro-gelatine cartridges, each weighing two ounces. There will thus be required about 100 lb. of the explosive for each 100 miles of the journey, and the cost, including a very liberal allowance for construction and insulation of the cartridges should not, he considers, exceed 1s. 8d. per lb. Three thousand pounds of fuel would, he modestly submits, thus more than provide for a transatlantic voyage, and the cost thereof should not exceed £200, which would be inexpensive for a vessel of its probable carrying capacity which, in addition to fuel and supplies, should easily, he says, transport 25 adults consisting of a crew of 10 and 15 passengers.

Dr. Pynchon does not append any details concerning the probable cost of such a machine or its aggregate weight. Being propelled through the air at the rate of 200 miles an hour by intermittent explosions of nitro-gelatine or dynamite will certainly be a unique and exhilarating experience, but we do not anticipate that those working at the problem will immediately abandon their theories for its application. When an aerial machine is devised that will cross the Atlantic in about 15 hours the Cunard Company's ocean Greyhound will, indeed, have to take a back seat!

DATE DUE

JUN 07 1984		
~~MAR 15 89~~		
FEB 1994		
GAYLORD		PRINTED IN U.S.A.